AN INTRODUCTION TO
LASER
SPECTROSCOPY

AN INTRODUCTION TO
LASER
SPECTROSCOPY

Edited by

DAVID L. ANDREWS AND
ANDREY A. DEMIDOV

University of East Anglia
Norwich, England

PLENUM PRESS • NEW YORK AND LONDON

Library of Congress Cataloging-in-Publication Data

An introduction to laser spectroscopy / edited by David L. Andrews and
Andrey A. Demidov.
    p. cm.
  Includes bibliographical references and index.
  ISBN 0-306-45203-0
  1. Laser spectroscopy.   I. Andrews, David L. II. Demidov, Andrey
A.
QC454.L3I595 1995   95-45655
                                              CIP

Incorporating proceedings of the Summer School in Laser Spectroscopy,
held September 4–10, 1994, in Norwich, England

ISBN 0-306-45203-0

© 1995 Plenum Press, New York
A Division of Plenum Publishing Corporation
233 Spring Street, New York, N. Y. 10013

10 9 8 7 6 5 4 3 2 1

Printed in the United States of America

PREFACE

In September 1994 a Summer School in Laser Spectroscopy was launched at the University of East Anglia in Norwich, under the auspices of the Engineering and Physical Sciences Research Council. The concept for this graduate school had been developed through extensive discussions within the chemistry and physics communities, supported by the Institute of Physics and the Royal Society of Chemistry. Drawing on the best existent university courses, the course was designed to be thoroughly cross-disciplinary, affording both breadth and depth in the subject of laser spectroscopy.

This book is one of the incidental fruits of the venture. Although representative of the lecture programme, it is not simply a volume of proceedings; for each chapter the authors here present an account specifically designed for accommodation into a free-standing graduate-level textbook. As with the Summer School, not only the research side of the subject is covered; significant areas of analytical and industrial applications are also included. As well as affording an overview of the subject, it is our hope that the book will allow newcomers to laser spectroscopy to reach a point where the many other more specialised reviews and research-level monographs become more accessible.

D.L. Andrews
A.A. Demidov
Norwich, June 1995

CONTENTS

SOURCES FOR LASER SPECTROSCOPY

Martin R. S. McCoustra

Department of Chemistry,
University of Nottingham,
Nottingham, NG7 2RD

Since the initial development of the laser in the early 1960's, almost all lasers have been applied in spectroscopic experiments of one form or another. Many of the laser sources first developed, such as the ruby laser and rare gas ion lasers, operate at a fixed frequency, providing only a single lasing wavelength, or at most lasing on a few narrow atomic or ionic resonances and hence tuneable only across the available optical cavity modes lying within the narrow gain profile of the relevant resonance (figure 1). As such, these lasers have proved difficult to employ spectroscopically, with the obvious exception of their application in Raman spectroscopy where excitation on a single narrow frequency band is obviously desirable. The true dawn of modern laser spectroscopy arose with the development of broadly tuneable lasers such as the organic molecular dye laser and it is upon such tuneable laser sources that the following discussion will focus. Spectroscopic measurements of one form or another can be made across the entire electromagnetic spectrum. However, the majority of studies and probably the most relevant to our current discussion are made in the region stretching from the microwave with wavelengths of around 1 cm to the vacuum

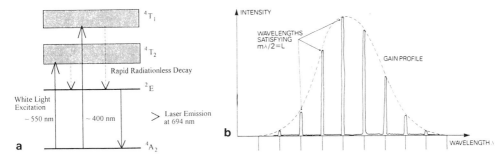

Figure 1. (a) Energetics of the ruby (Cr^{3+}:Al_2O_3) laser. (b) Emission spectrum of a typical laser. Emission occurs only at wavelengths within the fluorescence linewidth of the laser medium which experience gain and also satisfy the standing-wave condition (Andrews, 1990). [Reproduced with permission of Springer-Verlag.]

An Introduction to Laser Spectroscopy
Edited by David L. Andrews and Andrey A. Demidov, Plenum Press, New York, 1995

1

ultraviolet beyond 200 nm. Each region reflects a particular type of energy level transition and hence spectroscopy; rotational in the microwave and far-IR (*ca.* 1 cm - 50 µm), vibrational in the mid- and near-IR (*ca.* 50 - 1 µm) and electronic in the near-IR through to the VUV and beyond (*ca.* 1 µm - 100 nm). In each of these regions, traditional sources of tuneable radiation have existed for a significant length of time; for example microwave generators, monochromatised globars and Nernst filaments in the IR, and monochromatised incandescent and discharge lamps in the near-IR, visible and UV. Moreover such traditional, incoherent sources have proved their utility in many beautifully detailed spectroscopic studies to be found in the literature of both the pre- and post-laser eras. However, in the two regions of the electromagnetic spectrum that this discussion will now focus on, the introduction of tuneable, coherent radiation sources based on lasers has led to a rebirth of high resolution spectroscopy.

TUNEABLE COHERENT RADIATION SOURCES IN THE VISIBLE AND UV

The Dye Laser

The development of the organic molecular dye laser some 25 to 30 years ago marked the true birth of modern laser spectroscopy, offering the spectroscopist a truly continuously tuneable coherent laser light source. Today, as then, the majority of dye lasers rely upon a solution of an organic molecular dye in a suitable solvent to act as its gain medium, although recent advances have seen the introduction of both plastic-encapsulated dyes and sol-gel technology thin film dye-based systems. The variety of available dyes ensures that the entire spectrum from around 320 nm to beyond 1000 nm can in theory be covered by a dye laser (figure 2). The dye molecules themselves are complex, conjugated organic molecules often containing multiple fused aromatic ring systems, as illustrated by the example of Rhodamine B (figure 3a). Such species are chosen to ensure that the majority of the exciting radiation absorbed by the molecule is returned as *fluorescence emission* and the processes which deplete the yield of fluorescence, such as non-radiative decay, generation of triplet states and photochemical reactions, are minimised.

The potential ability of a molecule to act as a laser dye is characterised by the quantity known as the *fluorescence quantum yield*, Φ_f. Photophysically, this is defined as,

$$\Phi_f = \frac{\text{number of fluorescence photons emitted}}{\text{number of incident photons absorbed}}$$

and hence the closer the value to 1, the better the potential as a laser dye. It is possible to show that Φ_f can be defined in terms of the rates of fluorescence decay and those of competing processes. The nature of some of these competing processes is best illustrated by the simplified *Jablonski Diagram* in figure 4. Upon excitation of the dye molecule, an excited vibrational level within an excited singlet electronic state is produced. This state is rapidly relaxed by collisions with the surrounding solvent molecules, a process known as *collision-induced vibrational relaxation (CIVR)*, to yield a vibrationally cold, but electronically excited molecule. It is this initial process that is responsible for the near invariance of the emission spectrum of dye molecules with excitation wavelength (figure 3b). In a normal molecule, several physical processes are then open at this point, in addition to the possibility of chemical reactions involving this excited state,

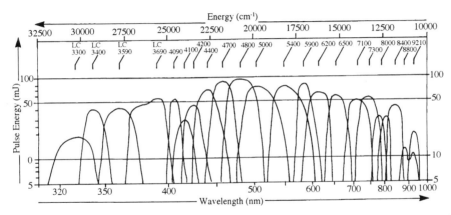

Figure 2. Tuning curves of a number of laser dyes across the near-UV, visible and near-IR when pumped by an exciplex laser at 308 nm. The LC designation is a manufacturer's catalogue number. [Reproduced with the permission of Lamba Physik GmbH.]

(i) The molecule can radiatively relax, *i.e.* undergo spontaneous emission. The excess energy then appears as a *red-shifted* fluorescence photon (process *iv* in figure 4). This process is characterised by a rate coefficient k_{rad} which is related to the spontaneous radiative lifetime of the excited state τ_{rad} by the expression,

$$k_{rad} = \frac{1}{\tau_{rad}} \quad .$$

(ii) Where an appropriate energy level exists, non-radiative transfer can occur to a nearby triplet state (process *iii* in figure 4). This process is known as *intersystem crossing (ISC)*. This is followed by more collisional relaxation until the lowest vibrational level in the triplet state is attained. Radiative relaxation can then occur by the spin-forbidden process of phosphorescence (process *v* in figure 4). This process can be characterised by a rate coefficient k_{tr} that incorporates details of the rates of ISC and triplet CIVR.

(iii) Again where an appropriate vibrational energy level exists, non-radiative transfer can occur to highly excited vibrational levels within the singlet ground electronic state. This is a process known as *internal conversion (IC)*. Such internal conversion is rapidly followed by collisional relaxation resulting in the completely non-radiative relaxation of the excited molecule (process *ii* in figure 4). Again a rate coefficient, k_{nr}, can be used to characterise this process.

Given the rate coefficients for the process described above, it is possible to show that we have the following relation;

$$\Phi_f = \frac{k_{rad}}{k_{rad} + k_{nr} + k_{tr}} \quad .$$

3

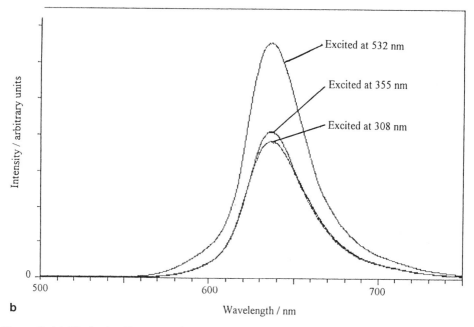

Figure 3. (a) Rhodamine B, an organic dye molecule commonly used as a laser dye: (b) Dispersed fluorescence emission curves of a dilute solution of Rhodamine B in methanol, excited at 308, 355 and 532 nm.

Hence, to maximise the fluorescence quantum yield, the rate coefficients for the competing radiative and non-radiative relaxation processes of the excited molecular singlet state must be minimised.

The radiative processes are, of course, governed by the *Franck-Condon principle*, resulting in emission occurring from the lowest vibrational level in the excited state (be it singlet or triplet) to a range of levels in the singlet ground state. The range of ground state vibrational levels accessible is largely governed by the overlap integrals or Franck-Condon factors for the transitions involved. It is this effect that gives the breadth to the fluorescence and phosphorescence spectra of such molecules and gives rise to the concept of the *vibronic laser* and the tunability of the organic molecular dye laser.

In a vibronic laser, such as an organic molecular dye laser, a combination of three- and four-level laser schemes is found to operate. In the three-level scheme (figure 5a), excitation occurs from the lowest vibrational levels in the ground singlet electronic state (level 1) to some excited vibrational level in the excited singlet electronic state (level 2), followed by collisional relaxation (CIVR) to the lowest vibrational level in that excited electronic state

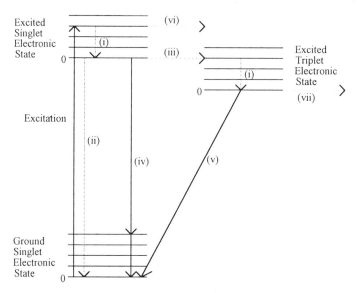

Figure 4. A Jablonski diagram showing the photophysical processes that can occur in a photoexcited molecule. (i) Collision-induced vibrational relaxation (CIVR): (ii) Internal conversion (IC) followed by CIVR on the ground electronic state: (iii) Intersystem crossing (ISC) followed by CIVR on the lowest triplet state: (iv) Fluorescence: (v) Phosphorescence: (vi) Reaction of excited singlet state: (vii) Reaction of triplet state.

(level 3). Spontaneous and stimulated emission can then occur between this level and level 1. In contrast, the four-level scheme reflects the rôle of the Franck-Condon principle in broadening the range of vibrational levels that are accessible in emission to the ground singlet electronic state. Levels 1, 2, and 3 in the four-level scheme (figure 5b) are as in the three-level scheme; however, spontaneous and stimulated emission from level 3 occurs to an excited vibrational level within the ground electronic state (level 4), which is then rapidly depopulated by collisional relaxation.

In operation, organic molecular dye lasers can be either pulsed or continuous wave (CW). At its simplest, the dye laser will consist of a simple optical cavity comprising a totally reflecting rear mirror, a suitable output coupler, and containing the gain medium as a flowing solution. For pulsed operation, relatively low flow rates are required and consequently simple liquid flow cells can be used to contain the gain medium. With CW operation, however, much higher flow rates are required to ensure continual replenishment of the dye solution within the excitation volume of the oscillator, and it has been found that liquid jet sources are ideally suited to this purpose. Such a simple laser will produce a coherent output beam with a bandwidth nearly as wide as the fluorescence emission spectrum of the dye itself (typically *ca.* 20 - 50 nm). To achieve narrow bandwidth operation with such a laser, additional wavelength-selective optical elements must be introduced into the oscillator cavity.

For narrow bandwidth pulsed operation, excitation of the gain medium can be achieved either through *flashlamp pumping* or by *laser pumping*. In the former, high intensity xenon discharge lamps are employed. These can provide multi-joule outputs in relatively long pulsewidths (from 10's of μs), which can potentially be compressed to nanosecond widths by electro-optic techniques, but are limited to a maximum repetition rate of a few Hz. The broad spectral output of these flashlamps means that energy is rather inefficiently converted

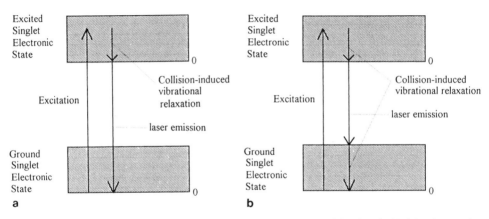

Figure 5. Energetics of an organic molecular dye laser showing (a) 3-level and (b) 4-level operating schemes.

from the input light into output laser light, with conversions of typically 5% - 10%. Furthermore, such flashlamps emit a considerable amount of short wavelength UV, which is inefficiently converted into visible laser output, and readily results in photochemical degradation of the dye. While such flashlamp-pumped dye lasers have their uses, especially in the medical field where the long pulsewidths reduce tissue damage, for spectroscopic purposes they have all but been superseded by laser pumping. Laser pumping with pulsed lasers offers a much more flexible choice of output pulse energies and pulse repetition rates, through appropriate choice of the pump laser as summarised in table 1. Moreover, laser pumping of pulsed dye lasers is generally somewhat more efficient in that the pump wavelength can be chosen to optimise the photophysics of the dye for laser action and to minimise photochemical degradation. For example, conversion efficiencies as high as 30% are not unusual for excitation of a rhodamine dye with the second harmonic (532 nm) output of a Nd^{3+}:YAG laser, and conversions of between 15% and 20% are not untypical for most pulsed laser-pumped dye lasers.

Table 1. Typical characteristics of laser pumped dye lasers and the more common laser pump sources.

Pump Laser	Dye Laser	
	Pulse Energy	Repetition Rate
Nitrogen (337 nm)	< mJ	< 10 Hz
Exciplex (XeCl 308 nm; XeF 351 nm)	< 50 mJ	< 500 Hz
Nd^{3+}:YAG (532 nm; 355 nm)	< 200 mJ	< 20 Hz
Copper Vapour (510.5 nm; 578.2 nm)	< mJ	< 20 kHz

For such pulsed dye lasers, two basic oscillator types have evolved as illustrated in figure 6. The *Littrow* dye laser, illustrated in figure 6a, uses either a telescopic or prism beam expander to fill the wavelength-selective optic, a diffraction grating, which is itself used as the cavity rear reflector. Operation of the grating in this configuration means that low grating orders are employed and the resultant laser bandwidth is somewhat broad at 0.2 - 2 cm^{-1} dependent on the detailed nature of the oscillator. Additional wavelength-selective optics, such as etalons, can be added to this type of cavity to further reduce the bandwidth to less than 0.1 cm^{-1}. In the *grazing incidence* design (figure 6b), light within the laser oscillator is incident upon the grating at near grazing angles and hence high grating orders. This naturally produces a somewhat narrower output bandwidth of typically less than 0.1 cm^{-1}, with the significant advantage of an optically simpler cavity.

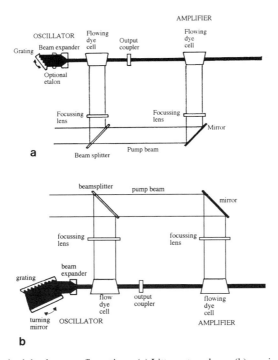

Figure 6. Common pulsed dye laser configurations: (a) Littrow-type laser; (b) grazing incidence dye laser.

With such pulsed dye lasers, the natural limit of the bandwidth is determined by the *transform limit* of the pulsewidth through the relationship,

$$\tau = \frac{0.441}{\Delta v} = \frac{0.441}{c\,\Delta\bar{v}} \quad,$$

where τ is the pulsewidth, Δv is the frequency bandwidth of the laser, $\Delta\bar{v}$ is the bandwidth in cm^{-1} and c is the speed of light in cm s^{-1}. Hence for a 10 ns pulsewidth, the narrowest bandwidth that can be achieved is of the order of 0.015 cm^{-1} (450 MHz). Both of the basic pulsed dye laser cavity designs can achieve such transform-limited operation by the addition of suitable bandwidth narrowing optics such as etalons. While such bandwidths are suitable for much spectroscopy, it is clear that narrower bandwidths can be achieved by increasing the pulsewidth of the laser. In the limit, the narrowest bandwidth operation will obviously be achieved using a CW dye laser.

As with pulsed dye laser cavities, the development of CW dye lasers has seen the evolution of two types of CW dye laser oscillator (figure 7). The *standing wave dye laser* most closely resembles the simple broadband oscillator described above. The addition of a simple birefringent filter performs the bandwidth-narrowing rôle and yields output bandwidths of the order of 0.1 cm^{-1} (3 GHz). Narrower linewidth operation can readily be achieved using an additional etalon to give bandwidths of *ca.* a few 10's of MHz. The alternative *ring dye laser* naturally provides in its basic form a much narrower bandwidth of *ca.* 0.01 cm^{-1} (300 MHz) and is capable of providing the ultimate in narrow bandwidths of 0.000015 cm^{-1} (450 kHz) when operated with additional line-narrowing optics. Normally such CW dye lasers are pumped using an argon ion laser, although excitation of blue and near-UV dyes is sometimes achieved using krypton ion lasers. Few, if any other CW lasers have the necessary output powers to excite a CW dye laser, nor is it possible to excite such a laser using a CW arc lamp. Conversion efficiencies are comparable to those that can be achieved using a pulsed dye laser.

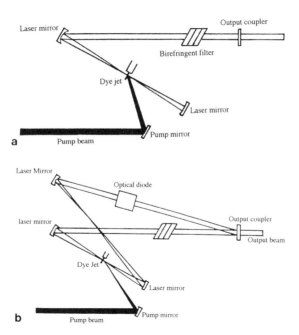

Figure 7. Continuous-wave dye laser configurations: (a) standing-wave and (b) ring dye laser configurations.

Transition Metal Ion Vibronic Lasers

Solid state laser systems based on transition metal or rare earth ions doped into an optically transparent crystalline solid have been with us since the dawn of the laser age. The first observation of laser action itself was made using such a transition metal ion laser; the ruby laser in which laser activity is associated with the chromium(III) ions doped into an aluminium oxide (Al_2O_3) matrix. The properties of this support medium are such that the ground $^4A_2$ electronic state of the Cr^{3+} ion exists as a single, narrow energy level. Consequently, laser action on the $^2E \rightarrow {}^4A_2$ transition results in emission at a single wavelength (figure 1a). Although the higher $^4T_2$ and $^4T_1$ electronic states of the Cr^{3+} ion are coupled to lattice vibrations and are consequently broadened into wide vibronic bands, these states play no active rôle in the laser system other than ensuring that population inversion in the ^2E level is possible, by rapid non-radiative relaxation within the vibronic bands.

By altering the support material from Al_2O_3 to $BeAl_2O_4$, the ground $^4A_2$ electronic state of the Cr^{3+} ion can then couple to the lattice vibrations of the support matrix and so form a wide vibronic band in its ground electronic state. Laser action in this material, known as alexandrite, then occurs both on a fixed wavelength $^2E \rightarrow ^4A_2$ transition at 680 nm (*cf.* the transition in ruby) and on a range of wavelengths from 701 to 826 nm coupling the $^4T_2$ vibronic state to the $^4A_2$ state (figure 8). In principle, the mechanism of laser action is essentially the same as that in operation in an organic molecular dye laser.

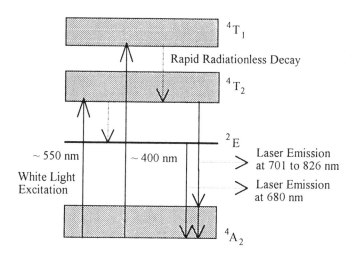

Figure 8. Energetics of the alexandrite (Cr^{3+}:$BeAl_2O_4$) laser.

In recent years, there has been a rapid development in such tuneable transition metal ion vibronic lasers. Many materials, some of which are listed in table 2, have been found to exhibit tuneable laser action either in pulsed mode, or CW, or even both, and the basic oscillator designs for such lasers are not too dissimilar from those employed with organic molecular dye lasers. These vibronic lasers, however, have some benefits over the traditional dye laser;

(**i**) Such lasers can provide higher output powers or pulse energies in regions of the spectrum traditionally difficult for dye lasers given the same energy input, *i.e.* they have higher conversion efficiencies. Perhaps the best example of this is the Ti^{3+}:sapphire laser which can provide as much as 2 to 3 times the output of a dye laser operating in the near-IR at around 800 to 900 nm.

(**ii**) The solid state materials are photochemically stable, removing the problem of photochemical degradation associated with organic molecular dye lasers.

(**iii**) When used with non-linear optical methods, tuneable transition metal ion lasers have the potential to provide coverage in the entire UV, visible and near-IR in a relatively compact all solid state source.

For all these advantages, the extremely wide coverage offered by dye lasers and the maturity of such devices as commercial products ensures that the dye laser remains today the most significant tuneable radiation source in the visible and UV.

Table 2. Types of metal ion tuneable vibronic lasers.

Laser Material	Pump Source	Operation	Wavelength Range / nm
Tm^{3+}:YAG	laser	CW	1870 - 2160
Co^{2+}:MgF_2	1320 nm Nd^{3+}:YAG	Pulsed	1750 - 2500
Cr^{3+}:Mg_2SiO_4 (forsterite)	laser	Pulsed/CW	1167 - 1345
Cr^{3+}:$LiSrAlF_6$	laser/lamp	Pulsed/CW	760 - 920
Cr^{3+}:$Be_3Al_2(SiO_3)_6$ (emerald)	laser	Pulsed/CW	720 - 842
Cr^{3+}:$LiCaAlF_6$	laser	Pulsed/CW	720 - 840
Cr^{3+}:$BeAl_2O_4$ (alexandrite)	laser/lamp	Pulsed/CW	701 - 858
Ti^{3+}:Al_2O_3 (sapphire)	laser/lamp	Pulsed/CW	660 - 1180
Ce^{3+}:YLF	KrF laser	Pulsed	309 - 325
Ce^{3+}:$LiSrAlF_6$	266 nm Nd^{3+}:YAG	Pulsed	285 - 297

Non-linear Optical Methods of Frequency Extension

Non-linear optical techniques allow the extension of the operating range of tuneable UV and visible lasers into the VUV and IR regions of the spectrum. The theory and practice of such non-linear optical methods will be dealt with in more detail in a subsequent chapter. Here, the implications of these methods will briefly be discussed.

The simplest of the non-linear optical methods is *second harmonic generation (SHG)*. By this technique, radiation of a frequency v and wavelength λ is introduced into a non-linear crystal, and radiation at a frequency $2v$ and wavelength $\lambda/2$ is generated therein. Standard wavelength separation techniques, such as dichroic mirrors or dispersive prisms, can be used to separate the co-propagating beams. Many materials have been shown to exhibit such non-linear responses, but much of the current interest is focused on materials such as β-phase barium borate (BBO) and the lithium analogues of this material. These inorganic, crystalline solids exhibit wider optical transparencies, down to *ca.* 190 nm for BBO and approaching 180 nm for the lithium borate materials, and higher non-linearities than more traditional non-linear materials such as potassium dihydrogen phosphate (KDP) and analogues thereof. Non-linear methods such as SHG have traditionally required the high intensity radiation associated with a pulsed laser. With the development of more efficient materials and the growing use of intracavity SHG methods, significant advances are being made in tuneable CW UV sources based on CW dye lasers and metal ion vibronic lasers.

Further extension into the deep-UV can be achieved using another non-linear technique; sum frequency generation (SFG). In this technique, two laser frequencies v_1 and v_2 are mixed in a non-linear crystal to produce the sum frequency v_{sum} ($=v_1 + v_2$). Typically one of the frequencies is fixed, for example the fundamental (1064 nm) or second harmonic (532 nm) of a Nd^{3+}:YAG laser, while the other is tuneable, for example the output of a dye laser or Ti^{3+}:sapphire laser or second harmonic thereof. If we consider the case of a Ti^{3+}:sapphire laser, the combination of frequency doubling and sum frequency generation can extend the accessible wavelengths from the fundamental range of *ca.* 700 - 1000 nm to cover most of the UV and visible as illustrated by table 3.

Closely related to the use of SFG is the coherent radiation source known as an *optical parametric oscillator (OPO)*. In an OPO, an intense pump beam is incident upon a non-linear crystal. Coherent radiation is then generated within the crystal at two frequencies known as the signal (ν_{signal}) and the idler (ν_{idler}), the sum of which produces the pump frequency ν_{pump} ($=\nu_{signal} + \nu_{idler}$). Construction of a wavelength-selective optical oscillator around the non-linear crystal will then result in the production of narrow bandwidth coherent radiation. While such devices have been known for some considerable time, it is only in re-

Table 3. Wavelength coverage of a hypothetical pulsed Ti^{3+}:sapphire laser pumped by the second harmonic of a Nd^{3+}:YAG laser.

Approximate Wavelength Range /nm	Mechanism
700 - 1000	Laser fundamental
416 - 506	Fundamental mixed with Nd^{3+}:YAG fundamental
350 - 500	Second harmonic
261 - 336	Second harmonic mixed with Nd^{3+}:YAG fundamental
234 - 334	Third harmonic
190 - 252	Third harmonic mixed with Nd^{3+}:YAG fundamental

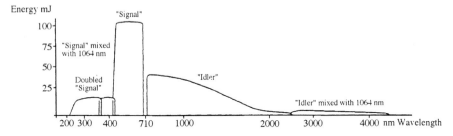

Figure 9. Performance of a commercial UV-pumped OPO. [Reproduced with the permission of Spectra-Physics, Inc.]

cent years with the introduction of more robust and efficient non-linear materials, such as BBO, that the OPO has been seen as a viable alternative to the dye laser and solid-state lasers, given the extremely wide output range possible from such devices. For example, a BBO OPO pumped by the third harmonic of a Nd^{3+}:YAG (355 nm) is capable of providing radiation from around 410 nm to beyond 200 nm. Doubling and sum frequency generation can extend this to around 200 nm, as shown in figure 9. While OPO generation has normally relied upon pulsed excitation, the higher non-linear efficiencies available from modern materials may see the growth in use of CW OPO sources.

The final non-linear technique to be considered neatly moves us into the IR region of the spectrum. Difference frequency generation (DFG) takes two frequencies v_1 and v_2, such that v_2 is less than v_1, and generates in the non-linear medium the difference frequency v_{diff} ($= v_1 - v_2$). While materials such as BBO can be used in the near-IR region, extension into the mid-IR requires complex IR-transparent material such as lithium niobate, silver gallium sulfide ($AgGaS_2$) and selenide ($AgGaSe_2$). With such techniques, the output of a tuneable dye laser or metal ion laser or of an OPO (see figure 9) can easily be extended beyond 1000 nm to at least 4000 nm. Specialist DFG techniques using multiple stages have demonstrated the possibility of generating radiation out to around 20 μm.

To conclude this section, it is pertinent to consider another frequency extension method, which while not a non-linear optical method in the same sense, is of considerable importance. *Stimulated Raman scattering* will be discussed in more detail later; suffice it to say that with an intense pulsed pump beam sufficient Raman scattering may be generated to stimulate further scattering on the same transition, enhancing the efficiency of the Raman effect and generating a beam of coherent radiation at the scattered frequency. Furthermore, if the stimulated Raman beam is itself of sufficient intensity, then further Raman scattering may occur on that wavelength. Stimulated Raman wavelength shifting has been shown to be a viable technique for the generation of both UV and IR tuneable radiation. The two methods most commonly employed are *stimulated vibrational Raman shifting* and *stimulated electronic Raman shifting* and each reflects the nature of the quantum states involved in the scattering process.

In the former, scattering is stimulated on vibrational transitions. A number of materials have been shown to be of use in this process, but by far the most widely used material is hydrogen. Both Stokes and anti-Stokes scattering can be stimulated, and for an input frequency v_{pump}, given the vibrational frequency of the hydrogen molecule $v_{hydrogen}$ (4155 cm⁻¹), scattered radiation may be generated at a number of frequencies,

$$v_{Stokes} = v_{pump} - nv_{hydrogen} \, , \qquad v_{anti\text{-}Stokes} = v_{pump} + nv_{hydrogen} \, ,$$

where n is known as the *order* of the scattering. This clearly offers the possibility of shifting the wavelength of an input beam by integral multiples of the vibrational frequency of the hydrogen molecule, both to the shorter wavelength side of the pump beam (anti-Stokes scattering) and to the long wavelength side (Stokes scattering). As an example, consider the

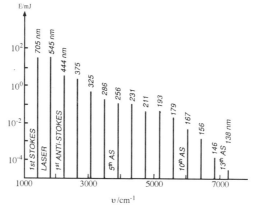

Figure 10. Measured energy of the first Stokes and up to the 13th anti-Stokes emission (138 nm) lines in H_2 gas (at 2-3 atm pressure) excited by 100 mJ laser radiation at 545 nm (Schomberg et al., 1983). [Reproduced with permission from H. Schomberg, H. F. Döbele and B. Rückle.]

data presented in figure 10. It is clear from these data, that for the first few orders of scattering, efficiencies of at least 1% are easily achievable and that extremely high orders of scattering are possible. In the case of Stokes scattering, it is worth noting that the efficiency of the process can be significantly increased if the pump radiation is focused into a silica tube, that acts as an optical waveguide, within the scattering medium.

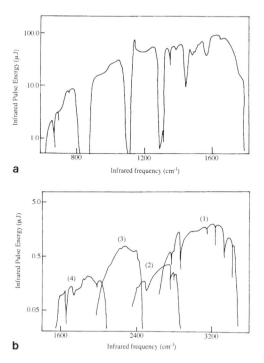

Figure 11. (a) Infrared tuning curve obtained by stimulated electronic Raman scattering in caesium vapour with DCM in the pump laser (vertical scale logarithmic). (b) Infrared tuning curves obtained by stimulated electronic Raman scattering in caesium vapour with rhodamine dyes in the pump laser (vertical scale logarithmic): (1) Rhodamine 590 in methanol (oscillator) and methanol/water (amplifier); (2) Rhodamine 610 in methanol; (3) Kiton Red in methanol/water; (4) Rhodamine 640 in methanol (Harris and Levinos, 1987). [Reproduced with permission from A. L. Harris and N. J. Levinos.]

In the latter, scattering is stimulated on an electronic transition within a gas phase alkali metal atom, for example the $6s$-$5d$ $^2D_{5/2}$ transition in caesium vapour (figure 11). This technique has been extensively used for the generation of tuneable IR radiation from tuneable visible sources. Conversion efficiencies are comparable to those for vibrational Raman shifting, at 1 to 10% for radiation output in the 1000 to 3000 cm^{-1} range, and operation using both nanosecond and picosecond pulses has been demonstrated. Such an IR source is considerably less complex than the corresponding difference frequency generation scheme required to reach similar mid-IR frequencies and it is significantly more efficient in producing tuneable coherent radiation.

TUNEABLE COHERENT RADIATION SOURCES IN THE IR

In contrast to the UV and visible regions, the IR suffers from a somewhat restricted choice of tuneable coherent radiation sources. Although recent developments in laser diode and non-linear optical technology promise a potential expansion in the number of tuneable IR sources, traditionally the choice of source has been limited to molecular gas lasers, lead salt diode lasers or colour centre lasers. In the following, each of the existing sources will be briefly discussed.

Molecular Gas Lasers

Amongst the earliest IR lasers to be developed and enjoying the unenviable position of being the most commercially successful of all lasers, the carbon dioxide (CO_2) lasers holds an important position in the laser community.   Operating on transitions between vibration-rotation energy levels in molecular CO_2, laser output in the region of 9.6 and 10.6 µm is on a series of closely spaced lines, at CW powers of up to several 100's of watts and pulse energies in the multi-joule level. This makes the CO_2 laser a *line-tuneable* as opposed to a continuously tuneable source.   Isotopic substitution can shift the output of the CO_2 laser somewhat, but again yielding line-tuneable radiation. If the pressure of the active medium (a nitrogen/carbon dioxide mixture) is increased, pressure broadening of the individual rovibrational lines can results in the formation of a quasi-continuum. However, the range of the output remains restricted to within a few 10's of cm^{-1} of the positions of the vibrational bands at 10.6 and 9.6 µm.

A number of other directly discharge-excited molecular gas lasers operating on molecular rovibrational transitions, such as the CO_2 laser, are available. The most important of these are the carbon monoxide (CO) laser which operates in the region 5.0 to 6.5 µm and the nitrous oxide (N_2O) laser operating in the 10.3 to 11.1 µm region. In contrast to the CO_2 laser, however, both of these lasers operate only in CW mode, and again are only line-tuneable. As in the case of CO_2, the potential exists for isotopic substitution and extension of the laser operating range.

For operation in the far-IR, beyond 40 µm, optically pumped molecular gas lasers are the source of choice. Material such as methanol (CH_3OH and isotopomers), methylamine (CH_3NH_2), formic acid (HCOOH) and methyl halides (CH_3X; X=F, Cl, Br, I) have all exhibited such behaviour. Laser activity requires that a narrow bandwidth pump laser (CO, CO_2 or N_2O) excites a single rovibrational transition in the gain medium, the low pressure vapour of one of the laser gases.   A population inversion is created between the excited vibrational state rotational energy levels and laser action results. Laser output is therefore again on a series of closely spaced lines dependant both on the nature of the gain medium and the vibrational mode within the molecule initially excited. Thus, for example, the most commonly used far-IR molecular gas laser medium, methanol, exhibits several groups of far-IR laser line at 70.6, 96.5, 118.8, 163.0, 570.5 and 699.5 µm.

While not offering the continuous tunability of other systems such as the dye laser or the lead salt diode laser (see below), the narrow bandwidth nature of these sources means that alternative spectroscopic techniques have been developed wherein a molecular resonance is brought into resonance with the effectively fixed frequency laser emission by electrostatic (Stark) or magnetic (Zeeman) field variations. Such Stark- and Zeeman-tuning have proved to be especially valuable where the large number of rotational or rovibrational lines available from a molecular gas laser at least allow some coarse tuning into a relevant molecular resonance.

Lead Salt Diode Lasers

Diode lasers are by far the most prolific of all types of laser.   Most of these lasers are made of semiconductor compounds of group III and group V species and have large bandgaps, with the result that emission occurs at wavelengths shorter than 2 µm. There is, however, a family of semiconductor materials containing elements mostly from groups IVb and VI of the periodic table which exhibit smaller bandgaps, corresponding to wavelengths in the 3 to 27 μm range. As most of these materials contain lead and exhibit a crystalline structure akin to that of rock salt, they are commonly referred to as *lead salt diode lasers*.

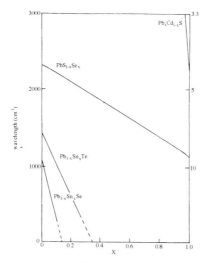

Figure 12. Dependence of the laser emission frequency on composition, at a temperature of 10 K, for Pb salts of various alloy compositions. [Reproduced from Demtröder (1982). Original source material, Spectra-Physics, Inc.]

The basic operating principle of such lead salt lasers is the same as that of the shorter wavelength diode lasers, as is also the case with the technology being applied to their growth. In principle, operation relies on the recombination of electrons from the semiconductor conduction band with holes in its valence band within the active region of a p-n diode structure. In contrast to the operation of a light emitting diode, the diode junction is buried in such a manner as to constrain emission within the p-n junction region, which is enclosed within appropriate optical surfaces, and hence to stimulate further emission within that region. Laser output then occurs from the edge of the diode structure and is continuously tuneable within a few 10's of cm⁻¹. Such diodes can be tuned in two manners. Extremely coarse tuning can be achieved by varying the composition of the diode material as shown in figure 12. Within an individual diode, fine tuning can be achieved by controlling the temperature of the diode and the current flowing through it (figure 13). Operation is CW and output powers are in the milliwatt range. The output is extremely narrow bandwidth, but (*cf* dye lasers) a large family of diodes of differing compositions is required to cover the entire mid-IR region.

One recent development in diode technology could promise a new IR source. The *quantum cascade laser* (see Faist et al., 1994) contrasts markedly with lead salt devices. Constructed as a multiple quantum well device from wide bandgap III-V materials, laser action occurs between energy levels located entirely within the conduction band of the semiconductor. The energy levels are determined by the dimensions of the quantum wells within the device. Laser action occurs at cryogenic temperatures, with outputs of 10's of milliwatts at 4.25 μm. This combination of mature and well-understood semiconductor technology with some unique laser device physics may promise both a reduction in the cost of IR laser diodes and their eventual operation at room temperature, as opposed to cryogenic temperatures.

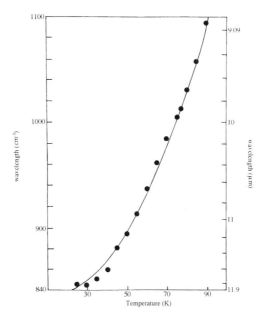

Figure 13. Temperature tuning of a Pb-Sn-Se laser. [Reproduced from Demtröder (1982). Original source material, Spectra-Physics, Inc.]

Colour Centre Lasers

In the near-IR region, between 1 and 4 μm, where many vibrational overtones and X-H stretching fundamentals are to be observed, the laser spectroscopist is offered a further potential source for experiments. When alkali halide crystals, such as sodium chloride

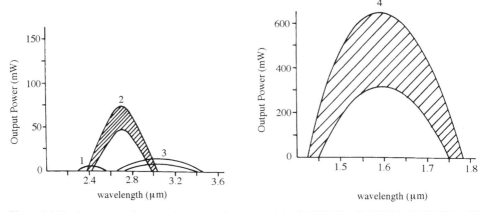

Figure 14. Tuning curves of some colour-centre laser materials: (1) KCl:Na; (2) KCl:Li; (3) RbCl:Li; (4) NaCl:OH. [Reproduced with the permission of Burleigh Instruments, Inc.]

(NaCl) and potassium chloride (KCl), are exposed either to high energy radiation (XUV light, X- and γ-rays) or when doped with excess alkali metal ions, crystal defects are formed which absorb visible radiation and are hence known as *colour centres*. These defects arise from the presence of vacancies surrounded by dopant ions, and contain free electrons. It is these electrons that give the defects their colour and hence their ability to act as a laser gain medium. When such a crystal is placed in a laser oscillator and excited by a CW ion laser,CW output with powers in the 10's and 100's of milliwatt range is obtained (figure 14). With oscillators normally of the standing wave design, additional optical line-narrowing elements are normally present to ensure that narrow bandwidth (*ca.* 1 MHz) operation is readily achieved.

REFERENCES

The following texts contain much information on the principles of laser action and on the applications of lasers in atomic and molecular spectroscopy. In addition, a series of articles entitled *Back to Basics* and published in the journal *Laser Focus World* contains much useful and easily acessible information on sources for laser spectroscopy.

Alves, A.C. ., Brown, J.M., and Hollas, J.M. (Editors), 1988, "Frontiers of the Laser Spectroscopy of Gases," Kluwer, Dordrecht.

Andrews, D.L., 1990, "Lasers in Chemistry, 2nd Edition," Springer-Verlag, Berlin.

Andrews, D.L. (Editor), 1992, "Applied Laser Spectroscopy," VCH, New York.

Demtröder, W., 1982, "Laser Spectroscopy: Basic Concepts and Instrumentation," Springer, Berlin.

Faist, J., Capasso, F., Sivco, D.L., Sirtori, C., Hutchinson, A.L., and Cho, A.Y., 1994, *Science*, 264:553.

Harris, A.L., and Levinos, N.J., 1987, *Appl. Opt.* 26:3996.

Hecht, J., 1992, "The Laser Guidebook, 2nd Edition," McGraw-Hill, New York.

Mollenauer, L.F., and White, J.C., 1987, "Tunable Lasers," Springer-Verlag, Berlin.

Schaefer, F.P., 1990, "Dye Lasers, 3rd Edition," Springer-Verlag, Berlin.

Schomberg, H., Döbele, H.F., and Rückle, B., 1983, *Appl. Phys. B* B30:131.

METHODS, DIAGNOSTICS AND INSTRUMENTATION

Martin R. S. McCoustra

Department of Chemistry
University of Nottingham
Nottingham, NG7 2RD

INTRODUCTION

While the source laser itself may be the core of a spectroscopic experiment, certain ancillary considerations are all but essential to ensure the success of the study. By far the most important of these is a knowledge of how much light the laser is emitting and the wavelength at which the emission is taking place. In the following, both of these very important parameters will be discussed in some depth.

THE MEASUREMENT OF OPTICAL POWER

Some Definitions

Prior to any detailed discussion of the various devices that may be used to measure optical powers, we must consider a few definitions. The first of these is the *irradiance* I, in W m^{-2}, as given by,

$$I = \frac{\text{Incident laser power}}{\text{Area irradiated}}.$$

For a CW laser, we can define the irradiance in terms of the power P, in watts, carried by the beam as,

$$I = \frac{P}{\pi r^2} = \frac{4P}{\pi d^2},$$

where r and d are the radius and diameter of the beam respectively. For a pulsed laser, we have a slight complication in that two types of irradiance can be considered. The first of these is the *average irradiance* I_{av} given by

An Introduction to Laser Spectroscopy
Edited by David L. Andrews and Andrey A. Demidov, Plenum Press, New York, 1995

$$I_{av} = \frac{ER}{\pi r^2} = \frac{4ER}{\pi d^2} \; ,$$

where E is the energy per pulse of the laser in joules and R is the laser repetition rate in Hz. The second is the *peak irradiance* I_{peak} and is given by the expression

$$I_{peak} = \frac{E}{\pi r^2 \tau_{pulse}} = \frac{4E}{\pi d^2 \tau_{pulse}} \; ,$$

where τ_{pulse} is the width of the laser pulse. It is worth noting that the laser power is generally not constant throughout a pulse but often varies, showing sharp, narrow spikes, as illustrated by the pulse temporal profiles in figure 1. The peak irradiance is therefore only a crude measure of the true variation of the laser power during a pulse. It may be that the instantaneous irradiance, at the peak of a sub-nanosecond spike within a nanosecond laser pulse, may exceed the 'peak' irradiance by many orders of magnitude. It is such spikes in irradiance that can cause significant damage to optical components.

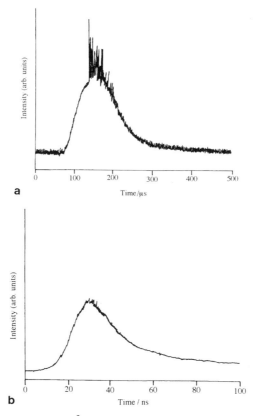

Figure 1. Temporal output of a pulsed Nd^{3+}:YAG laser operating at 1064 nm. (a) The output of the laser operating at fixed Q. (b) The output of the laser operating in a Q-switched mode. Note the sharp spikes present in both cases.

As a comparison of the irradiances available from lasers, it is worth remembering that the average solar irradiance is typically 1.5 kW m^{-2}, while a small 5 mW laser with a 1 mm diameter beam typical of those used for the alignment of optical systems delivers an irradiance of *ca.* 5 kW m^{-2}! A standard Nd^{3+}:YAG giving 800 mJ of 1064 nm output per pulse, (10 ns pulses at 10 Hz in a 6 mm diameter beam), typical of the lasers used to pump dye lasers, can deliver an average irradiance of *ca.* 3 MW m^{-2} while the peak irradiance is around 3 GW m^{-2}. The high degree of collimation of most laser beams also means that the irradiance is readily increased by focusing. If it is assumed that the laser beam has a *gaussian-mode* distribution of power across the beam, then we can focus that beam to the smallest possible spot size or the *diffraction limit*. The radius w_0 of this diffraction limited spot can be calculated either from the relationship,

$$ w_0 = \frac{f\lambda}{\pi r} , $$

where λ is the wavelength of the light, r is the radius of the laser beam on the lens of focal length f, or from the relationship

$$ w_0 = \frac{2M^2\lambda}{\pi\theta} , $$

where M^2 is a measure of the beam quality and θ is the angle of convergence of the focused beam. This latter relationship is of more practical use when the beam under consideration is not purely gaussian. The M^2 parameter (Johnston, 1990) as a measure of beam quality can be determined by measuring the distribution of power across the beam and takes a value of unity for a pure gaussian-mode beam. Values of M^2 less than 1.2 are normally thought to represent a high quality beam. If we now consider the examples of the laser beams discussed above focused with a 50 mm focal length lens to the diffraction limit, then for the He-Ne laser an irradiance of 11 MW m^{-2} is obtained. For the pulsed Nd^{3+}:YAG laser, the peak irradiance now exceeds 2,000 TW m^{-2}.

To conclude this brief definition of terms, the final concept to be considered is the photon flux I_{photon} in quanta m^{-2} s^{-1}. For a laser beam of wavelength λ , the photon flux is determined from the irradiance by the relationship

$$ I_{photon} = \frac{\text{Irradiance}}{\text{Photon Energy}} , $$

and hence,

$$ I_{photon} = \frac{I\lambda}{hc} , $$

which, taking our example laser beams gives *ca.* 2×10^{22} quanta m^{-2} s^{-1} and 9×10^{30} quanta m^{-2} s^{-1} for the unfocused He-Ne and Nd^{3+}:YAG laser respectively.

The measurement of optical power can now be discussed. Given the wide range of powers that it may be required to measure, a wide range of devices has been applied. However, these devices can be broadly classified into two groups. *Quantum detectors* rely on the quantum nature of light for their basic operation, while *thermal detectors* are based on the heating effects associated with high power laser beams.

Measurement of Optical Powers using Quantum Detectors

As stated above, quantum detectors rely for their operation upon the direct interaction of the photons in a laser or any other optical beam with the detector. We can divide these detectors into three basic types; *photoemissive, photovoltaic* and *photoconductive*. The latter two represent different modes of operation of basically the same photosensitive device.

Photoemissive devices are best represented by the *photomultiplier tube (PMT)* which represents a practical application of the photoelectric effect. When light is incident upon a surface in vacuum, provided that the energy of the photons ($h\nu = hc/\lambda$) is greater than that which binds electrons to the surface, E_b, then electrons will be emitted from the surface into the vacuum with a kinetic energy E_k given by,

$$E_k = h\nu - E_b = \frac{hc}{\lambda} - E_b \; .$$

By choosing a material in which the binding energy is small, electrons can be emitted at all wavelengths shorter than *ca.* 1000 nm. Such materials can then be used as the *photoemissive cathode* or *photocathode* in a PMT. A variety of these materials exist but the most commonly used are compounds of the alkali metals or alkali metal alloys with antimony. For example, the S11 photocathode is caesium antimonide (CsSb), the S20 trialkali photocathode is (NaKCs)Sb and the bialkali photocathode can be (KCs)Sb, (RbCs)Sb for normal temperature range operation or (NaK)Sb for high temperature operation. The efficiency with which a photocathode converts photons into electrons, known as the *quantum efficiency* η varies markedly both with the composition of the photocathode and with the wavelength of light incident on the photocathode, as shown in figure 2. In a typical PMT, as shown in figure 3, the photocathode is operated in transmission mode and electrons ejected from it are accelerated and focused on to the first of a series of *electron multiplier dynodes*. A single electron incident upon such a dynode will eject several electrons from it. Each of these *secondary electrons* is then accelerated on to the next dynode where more electrons are generated. In effect, the single photoelectron generates a cascade of secondary electrons in the dynode chain, *i.e.* the dynode chain produces internal amplification within the PMT. The amplification factor or *gain G* may be

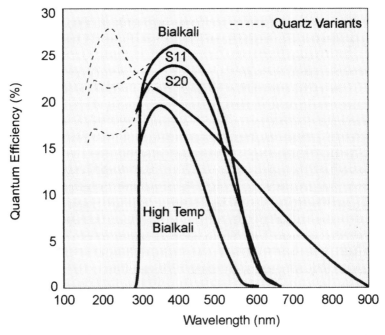

Figure 2. Typical response curves for 52 mm diameter photomultipliers. [Reproduced with the permission of Thorn EMI Electron Tubes Ltd.]

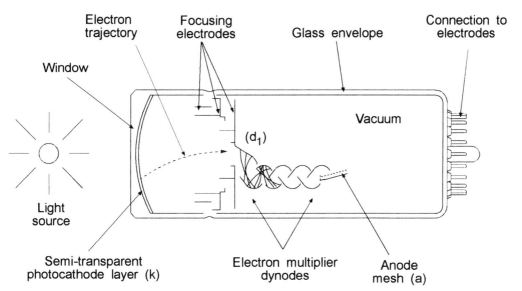

Figure 3. An illustration of the mechanism of operation of a photomultiplier tube. [Reproduced with the permission of Thorn EMI Electron Tubes Ltd.]

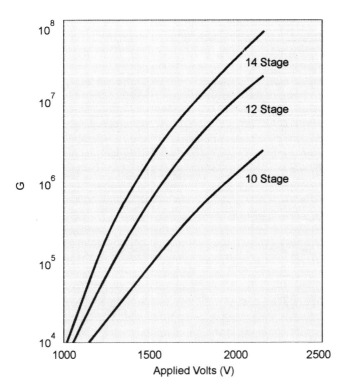

Figure 4. The variation of photomultiplier tube gain G with applied voltage, illustrating the effect of the number of dynode stages. [Reproduced with the permission of Thorn EMI Electron Tubes Ltd.]

as high as 10^6 or 10^8 in a well-designed PMT, but it is strongly dependent on the accelerating voltages within the tube, as shown in figure 4. The output of the dynode chain is collected on the anode and the signal appropriately sensed.

From the quantum efficiency of the PMT, the *radiant sensitivity* E in A W^{-1} is easily determined from the relation,

$$E = 8.06 \times 10^{-4} \lambda \eta(\lambda) \, ,$$

where λ is the wavelength of the radiation in nm. For a light source of optical power P in watts, the photocurrent generated at the photocathode i_c in amps is given by

$$i_c = 8.06 \times 10^{-4} \lambda \eta(\lambda) P \, ,$$

and hence the anode current i_a is

$$i_a = 8.06 \times 10^{-4} \lambda \eta(\lambda) PG .$$

While valid at low optical powers, high levels of illumination can lead to *saturation, i.e.* the output of the photocathode reaches a limiting value, and ultimately to damage to the photoemissive surface. The limiting current naturally depends on the type of photocathode material as demonstrated in table 1, and is dependent on the area of material illuminated. Illuminating a small area of a photocathode with an intense light source will therefore result in the generation of the limiting current, as opposed to that which might be expected from the radiant sensitivity. It might also lead to destruction of the photocathode material. For this reason, PMT's are rarely used to directly measure laser powers and are normally restricted to low intensity detection such as in laser-induced fluorescence.

Table 1. Photocathode limiting currents.

Photocathode Material	Limiting Current / nA cm^{-2}
Bialkali	2.5
High Temperature Bialkali	2.5
S11	15
S20	250

As an example of such an application, consider the detection of a 500 nm light source with an optical power of one photon per second (*ca.* 4×10^{-19} W). If incident upon a photocathode of quantum efficiency 20%, the photocathode current generated is of the order of 3×10^{-18} A. With a gain of 10^6, not untypical of a good PMT, the anode current would therefore be 3×10^{-12} A. Such a current is well within the capabilities of modern current sensing devices.

To conclude this discussion of PMT's, it is worth briefly considering the main sources of background signal and noise within these devices. The major source of noise is associated with the quantum statistical fluctuations in the number of photoelectrons emitted by the photocathode (*photoelectron noise*) and in the number of secondary electrons emitted by the dynodes (*multiplication noise*). Neither of these can be controlled by the user and are inherent in the quantum nature of the physical processes involved. In contrast, background and noise arising from the thermal emission of electrons from the photocathode and dynodes, and from radiation-induced emission, can be controlled to some extent. Thermionic emission is readily reduced by cooling the PMT, while selection of radioisotope-free materials in the construction of the PMT can reduce or eliminate the latter.

The remaining two types of quantum detectors rely on the same physical device, the *photodiode*. In operation, light incident on a semiconducting junction results in the generation of charge carriers; electrons in the conduction band of the device and holes in the valence band, (*cf.* the laser diode in the previous chapter). In *photovoltaic* mode, preferred for continuous and low frequency (< 1 kHz) detection, the photogenerated charge carriers are allowed to diffuse to the metal contacts on the diode and generate a photocurrent i_d, given by;

$$i_d = 8.06 \times 10^{-4} \lambda \eta(\lambda) P,$$

which is sensed. In *photoconductive* mode, favoured for high frequency and pulsed detection, a reverse bias voltage is applied across the diode and the photogenerated conductors increase the conductivity of the device. The increase in current flow is then measured.

The main advantages of these devices over PMT's are their much high quantum efficiencies and their extended operating range (figure 5). Devices based on silicon readily

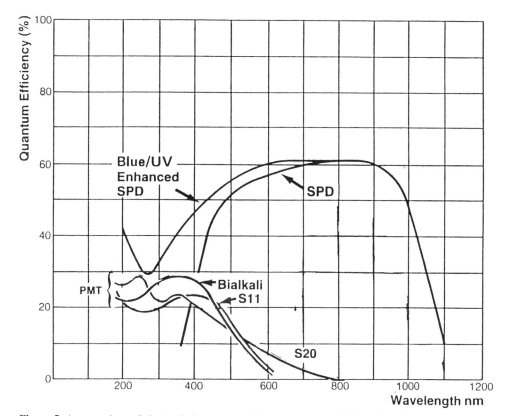

Figure 5. A comparison of photocathode response with the response of silicon photodiodes. [Reproduced with the permission of Thorn EMI Electron Tubes Ltd.]

achieve quantum efficiencies in excess of 60% and are able to operate beyond 1.1 μm. Devices based on other materials can extend operation into the near-IR (GaAs, InSb) and even into the mid-IR (CdHgTe). For all these advantages, such devices suffer significantly in terms of overall sensitivity when compared to PMT's. As is obvious from a comparison of the expressions given above for the photocurrent generated in a photovoltaic device and for the anode current in a PMT, for similar quantum efficiencies the photovoltaic device will produce G times less current. External electronic amplification, with its associated problems of additional noise, is then required to raise the photovoltaic device output to a measurable level. However, this lessened sensitivity does mean that such devices can be operated at much higher incident optical powers that a PMT.

Measurement of Optical Powers using Thermal Detectors

In contrast to a quantum detector, thermal detectors rely on the heating effects associated with the interaction of light with matter. One immediate consequence of this is that the wavelength-dependent response of a quantum detector is lost and therefore most thermal detectors exhibit an identical response whatever the wavelength incident upon the device. Two basic types of thermal detector are currently in use; the *pyroelectric detector* and the *calorimeter*.

Pyroelectric devices employ special crystalline materials that can directly convert into an electrical signal the thermal surge associated with an optical pulse. To measure CW radiation, therefore, that beam must be modulated to yield a pseudo-pulsed signal. The materials used, such as lithium tantalate ($LiTaO_3$), exhibit the piezoelectric effect and when distorted by a thermal pulse produce an electrical signal. Operation with such devices from a few microwatts to a maximum of around a few watts is possible, with the upper limit being imposed by the thermal damage threshold of the sensor material. As such these devices fill the gap in power/energy measurement between quantum devices and calorimeters.

In contrast to the more or less direct measurement of optical power by a pyroelectric detector, a calorimeter offers an indirect means of measuring such powers. At its simplest a calorimeter consists of an absorbing material which is heated by the laser beam. The temperature change in the absorber is monitored by a suitable temperature sensor such as a thermocouple or thermistor. The ideal calorimeter comprises the absorbing medium, which may be an optical glass filter or a graphite coating on a metal substrate, in good thermal contact with a heatsink. Taking this simple model, it is possible to investigate the dependence of the calorimeter sensitivity on detector parameters such as heat capacities and thermal conductivities by assuming that the radiation absorbed by the element raises its temperature, in turn leading to heat transfer to the sink. The rate at which energy is deposited in the absorbing element depends on the incident optical power P and the fraction of this radiation absorbed by the absorbing element η. If it is assumed that the absorber has a mass m_a, specific heat capacity c_a in J kg^{-1} K^{-1} and that the thermal conductivity of the link between the absorber and heatsink is G, then the temporal dependence of the absorber temperature T_a is given by;

$$\eta P = m_a c_a \frac{dT_a}{dt} + G(T_a - T_{surr}),$$

where T_{surr} is the (constant) temperature of the surroundings. For a constant optical power, a steady state will be achieved such that,

$$\frac{dT_a}{dt} = 0 \,,$$

and hence it is possible to show that

$$T_a = \frac{\eta P}{G} + T_{surr} \,,$$

i.e. the final temperature of the absorber is independent of its heat capacity and depends only on the rate at which heat is lost to the heatsink and surroundings.

In a more general case, however, the optical power P will be time-dependent. If it is assumed to show a periodic dependence of the form,

$$P = P_0 (1 + A \, \cos\omega t),$$

where ω is the modulation frequency, then it is possible to show that

$$T_a(\omega) = T_{surr} + \Delta T \cos(\omega t + \phi) \,,$$

where

$$\Delta T = \frac{\eta P_0 A}{\sqrt{G^2 + \omega^2 m_a^2 c_a^2}} \,.$$

Thus the temperature of the absorbing element exhibits a similar modulation to that of the optical power, but with a phase shift ϕ determined by,

$$\tan(\phi) = \frac{\omega \, m_a c_a}{G} \,.$$

This phase shift determines the response time of the calorimeter. For optimum performance, it is desirable to reduce this as much as possible, and this is best achieved by reducing the heat capacity of the absorber $m_a c_a$. Furthermore, the sensitivity of the calorimeter can be described in terms of the ratio $\Delta T/P_0$ and this is obviously maximised if both the heat capacity of the absorber and the thermal conductivity of the link to the heatsink and surroundings are as small as possible. A small absorber heat capacity is therefore a highly desirable trait in the calorimeter design.

For detection of pulsed optical power, the absorbing element must be able to integrate the absorbed power over the period of the optical pulse. If the pulsewidth τ is shorter than the timescale in which thermal transfer occurs to the heatsink and surroundings then the energy transfer term can be neglected and it is found that,

$$\Delta T = \frac{1}{m_a c_a} \int_0^\tau \eta P \, dt,$$

where the integral represents the absorbed power. In this regime ΔT becomes dependent only on the heat capacity of the absorbing element, and can be maximised by minimising the heat capacity. Thus, irradiating a glass filter of area A, thickness d and density ρ with a rectangular optical pulse of power P_{peak} and width τ will result in a temperature rise ΔT given by,

$$\Delta T = \frac{\eta P_{peak} \tau}{Ad\rho c_a}.$$

Taking as an example a 800 mJ pulse of 10 ns width incident upon a fully absorbing crown glass filter ($\rho = 2600$ kg m^{-3}, $c_a = 670$ J kg^{-1} K^{-1}) of 5 mm thickness and 6 mm in radius, the resulting temperature rise is estimated to be of the order of 0.8 K! Such a temperature rise is easily sensed.

It is clear from the above that to maximise ΔT, the thickness of the absorber must be reduced. This gives rise to a useful division between calorimeters. Those with a significant thickness of absorbing material, *e.g.* using a 5 mm glass filter, are known as *volume absorbers*. As a result of the large thermal mass of the absorber, these devices tend to be rather slow to respond but are readily capable of withstanding optical powers in the kilowatt range. At the other extreme are the *surface absorbers*, based for example on a graphite film, where the thickness of the absorbing material has been reduced to a few μm. While exhibiting a fast response due to their low thermal mass, high power pulses can readily evaporate the absorbing film. The use of surface absorbers is therefore usually restricted to a maximum power of few watts. However, such devices are readily capable of sensing optical powers in the micro- and milliwatt ranges.

THE MEASUREMENT OF LASER WAVELENGTH

With tuneable lasers capable of operating throughout the entire UV, visible and IR at bandwidths of better than 0.1 cm^{-1}, it is clear that laser wavelength is a critical experimental parameter that requires accurate and precise measurement. There are essentially two methods of determining the wavelength of a laser, *spectroscopic calibration* and *instrumental measurement*, and each of these will now be considered in turn.

Spectroscopic Calibration of Laser Wavelength

The recording of standard spectra provides perhaps the simplest means for the measurement of laser wavelength. In each of the spectroscopic regions discussed in the preceding chapter, workers have employed measurements of direct absorption and/or the emission spectra of simple, well-characterised systems that have themselves been compared previously with primary wavelength standards and can then be regarded as secondary standards. The choice of species depends entirely on the region of the spectrum under study, but in general either atoms or simple, stable diatomic or triatomics are employed.

The choice of secondary standards in the IR is relatively simple. In the far-IR, the pure rotational spectra of species such as CO, HCl and HF have found widespread use. Similarly in the mid-IR, the readily resolvable rovibrational bands of the above and of CO_2 and N_2O and their isotopically substituted analogues are frequently employed.

In the visible and UV, the situation is somewhat more complex. In emission studies, the dispersed spectra of iron and uranium hollow cathode lamps are a recognised standard. However, these are obviously difficult to directly relate to an absorption experiment based on a tuneable laser. One solution to this is to employ the spectroscopic technique known as *optogalvanic spectroscopy (OGS)*. In a hollow cathode discharge, a distribution of excited state atoms lying near the ionisation threshold is produced, and a steady state ion current is generated by a combination of collisional and field ionisation of the excited atoms. When a tuneable laser light source is introduced into such a discharge, photoexcitation of the excited atoms results in a change in the ion current within the discharge. The change can either be an increase, as excited atoms are photoionised, or a decrease as excited atoms are stimulated to relax. The spectra observed can then be related to the known emission of the discharge gas. Both iron and uranium have been employed in this manner. Moreover, the common buffer gases used in hollow cathode lamps, argon and neon, have well-known optogalvanic spectra (figure 6). The spectra obtained from OGS studies are relatively simple consisting of a number of narrow features. Such spectra contrast markedly with the alternative secondary standards in the visible and UV, the absorption spectra of molecules such as I_2 and Te_2. These spectra which have been recorded at the Doppler limit using Fourier transform methods, show extremely complex rovibronic structure. Consequently, it is often difficult to compare coarsely structured spectra of standards recorded at degraded resolution with these finely structured, Doppler-limited spectra. However, while OGS data are ideal for calibration over an extended range of wavelengths, there is no doubt that the I_2 or Te_2 absorption spectra are ideal for studies involving a limited range of wavelengths at Doppler-limited or sub-Doppler bandwidths.

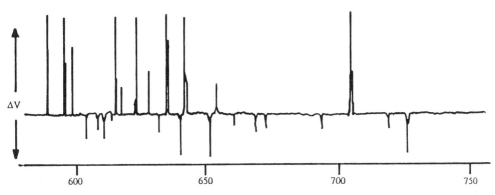

Figure 6. Complete first-order optogalvanic spectrum of a neon discharge in the visible region. Vertical scale is relative voltage deviation from the quiescent value. Horizontal scale is the laser wavelength in nanometres (Nestor, 1982).

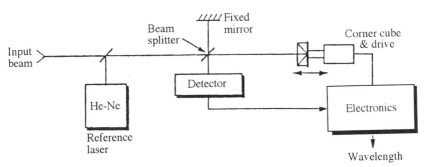

Figure 7. Schematic representation of a scanning Michelson interferometer as used for laser wavelength measurement (Cotnoir, 1989). [Reproduced with permission from L. J. Cotnoir]

Instrumental Measurement of Laser Wavelength

Instrumental methods of wavelength determination are well reported in the literature (Cotnoir, 1989; Synder and Hansch, 1990) and can be divided into two groups; the *dispersive methods* and the *interferometric methods*. Dispersive methods are based on traditional prism or diffraction grating monochromators to measure the laser wavelength. While used widely in laser laboratories, such devices have a number of drawbacks and are usually limited to applications in which wavelength needs to be determined to only a few tenths of a nanometre, although determination to within a few hundredths of a nanometre is feasible with larger instruments. The major drawbacks are threefold. The first, and perhaps the most important, is that the high powers of CW and pulsed lasers can result in optical damage to the prisms or grating within such instruments. Diffraction gratings are particularly susceptible to damage. To solve this, and also coincidentally the second drawback of dispersive instruments, wavelength inaccuracies associated with incomplete illumination of the dispersive element by a narrow collimated light source, a ground glass or silica plate is used to diffuse and scatter the collimated radiation before imaging onto the monochromator entrance slit. This technique both reduces the intensity of the light reaching the dispersive element and ensures that it is fully illuminated.

The final and most obvious problem with monochromators is the limited resolution achievable with such devices for a given set of physical dimensions. For any dispersive instrument the *resolving power R* is given by,

$$R = \left| \frac{\lambda}{\Delta\lambda} \right| = \left| \frac{\nu}{\Delta\nu} \right| = \left| \frac{\overline{\nu}}{\Delta\overline{\nu}} \right| ,$$

where $\Delta\lambda$, $\Delta\nu$ and $\Delta\overline{\nu}$ are the minimum separations of the line centres of two closely spaced lines that are just resolved according to the *Rayleigh criterion*. In effect $\Delta\lambda$, $\Delta\nu$ and $\Delta\overline{\nu}$ can be considered to be the laser bandwidth in appropriate units. Even for a pulsed laser of modest bandwidth, R must of necessity be extremely large, *e.g.* for a 500 nm beam with bandwidth of 0.05 cm^{-1}, (not untypical of a narrow band dye laser), a resolving power of 400,000 will be required to ensure that the wavelength of 500 nm is resolved from one of 500 nm $\pm$ 0.05 cm^{-1}.

The actual resolving power of a particular instrument depends on the dispersive element employed. For a prism monochromator, it can shown that,

$$R = g \frac{dn}{d\lambda} ,$$

where g is the length of the prism base and $dn/d\lambda$ is the dispersion of the prism material. Clearly to increase R, the size of the prism and/or the dispersion of the prism material can be increased. The latter increase occurs near the absorption edge of the material; thus silica is ideal in the 180 - 350 nm region while glass is more appropriate to the red of this region. Although for the silica prism the dispersion is greatest in the 180 - 350 nm region, significant absorption begins to occur at the shorter wavelength end of the region. This absorption will be further enhanced if physically large prisms are used to maximise the resolving power. Such absorption can result in damage to the prism on illumination with light from a pulsed laser. As a consequence of this and their significant cost, large prism-based instruments are not favoured for laser wavelength determination.

The alternative is a diffraction grating. In this case the resolving power is given by

$$R = mN,$$

where m is the order of the diffraction and N is the total number of grooves illuminated. Thus increasing m and/or N will result in an increase in resolving power. Obviously, however, there are practical limits to the density with which grooves can be ruled on a grating and there are also limits on the physical size of such grating set by considerations of construction cost. Thus while it may potentially be possible to construct a grating monochromator with the required resolving power, in practice the costs of such an instrument and the other factors discussed above limit the use of grating monochromators to devices that can localise wavelength to within a few tens or hundredths of a nanometre.

The alternative to a dispersive method is an interferometric technique, of which there are three commonly employed in laser wavelength determination. By far the most widely employed in both the UV/visible and in the IR is based on the *Michelson interferometer* (figure 7). Light from the laser is split by a beamsplitter positioned between two paths of nearly equal optical length. One path has a fixed mirror or corner cube, while the other has a mirror or corner cube that is driven by a motorised translation stage. When the beams are recombined on a simple photodetector, interference produces a sinusoidally varying signal with a peak each half-integer wavelength traversed by the moving mirror or corner cube. This signal can be described by the expression,

$$I(t) = \tfrac{1}{2}I_0 \left[1 - \sin(2\pi v/\lambda)\right],$$

where v is the velocity of the moving component. The modulation frequency of the interferometer signal is thus,

$$f = \frac{2\pi v}{\lambda} \ .$$

If the velocity of the moving component is fixed for both the unknown wavelength and for a reference laser wavelength, then the unknown is simply obtained from the ratio of the two modulation frequencies. Typically, the reference laser is a helium-neon laser and it is the frequency stability of this laser that limits the accuracy with which another wavelength can be determined. Off-the-shelf standard He-Ne lasers allow an accuracy of the order of 1 part in 10^6, *i.e.* a resolving power of 10^6, while frequency stabilised devices can offer 1 part in 10^8 accuracy.

Although ideal for CW laser wavelength measurement, a Michelson interferometer is impractical for use with pulsed lasers and alternative methods based on Fabry-Perot and Fizeau interferometers have evolved (Reiser, 1988). The basic *Fabry-Perot interferometer* consists of two plane-parallel, loss-free surfaces separated by an air space of width d. If the reflectivity of the plane surfaces is r, then the transmitted intensity is given by,

$$I(m) = \frac{I_0}{1 + \left[4r/(1-r)^2\right]\sin^2(\pi m)} \ ,$$

where

$$m = \frac{2d\cos\theta}{\lambda} \ ,$$

and θ is the angle of incidence of the light on the interferometer. Such a device can be employed either in a scanning mode in which d is mechanically varied or as a static device. In the former case, the transmitted intensity is measured as a function of d at fixed θ, and often displayed on an oscilloscope synchronised to the scanning. This approach is ideal for

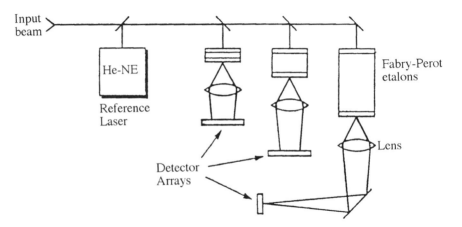

Figure 8. Schematic representation of a device using multiple fixed Fabry-Perot interferometers, as used for laser wavelength measurement (Cotnoir, 1989). [Reproduced with permission from L. J. Cotnoir]

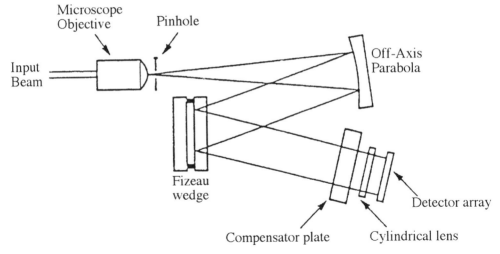

Figure 9. Schematic representation of a device using a Fizeau interferometer, as used for laser wavelength measurement (Cotnoir, 1989). [Reproduced with permission from L. J. Cotnoir]

displaying the mode structure of a narrow bandwidth CW laser, but of little practical use for pulsed lasers.

The static configuration is more applicable to pulsed lasers. Here, the Fabry-Perot interferometer with fixed d is illuminated at a range of angles θ using a converging or diverging laser beam. The result is a series of ring-like fringes which can be sensed by a suitable array detector. For a lens of focal length f, the pth fringe is found to have a diameter D,

$$D = 2f \left(2 - \frac{p\lambda}{d}\right)^{1/2}.$$

By sampling the fringe pattern in an appropriate manner, the laser wavelength can be estimated from the measured fringe diameter and the physical parameters of the device, f and d. Comparison of an unknown laser wavelength with that of a reference is thus easy provided that the two wavelengths share the same integral component of p. This is obviously true if the two wavelengths are very close to each other, *i.e.* within the *free spectral range FSR* ($=\lambda^2/d$) of the Fabry-Perot interferometer, but not so for distinctly different wavelengths. Clearly, the smaller the value of d, the larger the FSR of the Fabry-Perot interferometer and consequently the wider the range of wavelengths that can be determined, but with reduced precision. The approach commonly employed is therefore to use more than one Fabry-Perot interferometer, as in figure 8, to provide the necessary accuracy and precision in wavelength measurement. A narrow Fabry-Perot interferometer with a large FSR or a monochromator provides the initial wavelength estimate, further Fabry-Perot interferometers of decreasing FSR being used to attain the desired precision in measurement. Such multiple Fabry-Perot devices are capable of accuracies of 1 part in 10^6, comparable to that of Michelson interferometric methods.

The final type of interferometer in common use is the *Fizeau interferometer*. This device, as shown in figure 9, consists of two uncoated plates with a spacing of *ca.* 1 mm, wedged at an angle of *ca.* 1 mrad. When illuminated by a collimated light beam, interference within the wedge gives rise to a set of straight, evenly spaced fringes which are ideal for detection by an array photodetector. The intensity of the fringe pattern varies with the distance along the wedge thus,

$$I(x) = \frac{I_0}{2}\left\{1 + \cos\left[2\pi\left(fx + \frac{m}{2}\right)\right]\right\},$$

where f is the spatial frequency of the pattern. Such devices have a comparable accuracy (1 part in 10^6) to that of the other interferometric methods, although they are somewhat simpler in construction.

REFERENCES

The following texts contain much useful information on the measurement of optical power, laser wavelengths and other important operating parameters of lasers. In addition, much useful information on photodetection is to be found in literature from manufacturers such as Thorn EMI Electron Tubes Ltd. and Hamamatsu. In particular, the Thorn EMI Photodetection Information Service publishes an extremely useful series of documents on the principles, performance and use of photomultiplier tubes.

Andrews, D. L. (ed.), 1992, "Applied Laser Spectroscopy," VCH, New York.
Cotnoir, L. J., 1989, *Laser Focus World* 25:109.
Demtröder, W., 1982, "Laser Spectroscopy: Basic Concepts and Instrumentation," Springer, Berlin.
Johnston, T. J., 1990, *Laser Focus World* 26:173
Nestor, J. R., 1982, *Appl. Opt.* 21:41.
Reiser, C., 1988, *Proc. SPIE* 912:214.
Synder, J. J., and Hansch, T. W., 1990, *Laser Focus World* 26:69.

ABSORPTION AND FLUORESCENCE

Michael N.R. Ashfold

School of Chemistry
University of Bristol
Bristol, BS8 1TS
U.K.

ABSORPTION, INDUCED AND SPONTANEOUS EMISSION

Consider two non-degenerate levels of an atom (or molecule) with respective energies E_1 and E_2 in the presence of radiation of a frequency satisfying the Bohr condition

$$\nu = (E_2 - E_1)/h\nu \ , \tag{1}$$

where h is Planck's constant, 6.626×10^{-34} J s. If an atom in the lower energy state $|1\rangle$ absorbs a photon of frequency ν it may be excited to the upper state $|2\rangle$. This process is termed *induced absorption* or, more generally, simply *absorption*. The probability per second that an atom will absorb a photon, dP_{12}/dt, is proportional to the number of photons of energy $h\nu$ per unit volume, $\rho(\nu)$, and is usually expressed as:

$$\frac{dP_{12}}{dt} = B_{12}\rho(\nu) \ , \tag{2}$$

where the proportionality constant B_{12} is the *Einstein coefficient of induced absorption* (units $J^{-1} m^3 s^{-2}$).

The radiation field can also induce an atom in the higher energy state to make the transition $|2\rangle \rightarrow |1\rangle$, with simultaneous emission of a photon of frequency ν. This process is referred to as *stimulated emission*. Its probability per second, dP_{21}/dt, may be written

$$\frac{dP_{21}}{dt} = B_{21}\rho(\nu) \ , \tag{3}$$

where B_{21}, the *Einstein coefficient of induced* (or *stimulated*) *emission*, is equal to B_{12} in the case that the levels $|1\rangle$ and $|2\rangle$ have the same degeneracy. Stimulated emission is sometimes referred to as a *coherent* process. The stimulated photon is emitted with the same frequency and the same wavevector (**k**) as the photon which induced the transition.

An Introduction to Laser Spectroscopy
Edited by David L. Andrews and Andrey A. Demidov, Plenum Press, New York, 1995

This, however, is not the only way in which an excited atom in state $|2\rangle$ may lose energy. It can also *spontaneously* convert its excitation energy into an emitted photon hv. Spontaneous emission is an *incoherent* process. It does not depend on the presence of a radiation field. For a given atom in a given state its probability per second is a constant, the *Einstein coefficient for spontaneous emission*, A_{21} (units s^{-1}).

The *natural* (or *spontaneous*) *radiative lifetime*, τ, of the excited state $|2\rangle$ is just the inverse of A_{21}. The parameters A_{21} and B_{21} are each related to the transition moment $\langle 2|\mu|1\rangle$ as shown in Eqn. (4):

$$A_{21} = \frac{8\pi h v^3}{c^3} B_{21} = \frac{16\pi^3 v^3}{3\varepsilon_0 hc^3}\langle 2|\mu|1\rangle^2 \quad , \tag{4}$$

where ε_0 is the permittivity of free space, 8.854×10^{-12} F m^{-1}, and c is the speed of light *in vacuo*. Since A_{21} scales as v^3, spontaneous emission processes will tend to be very important at high frequencies, e.g. in the visible and ultraviolet spectral regions, but of negligible importance in the microwave and radio frequency regions. This accounts for the difficulties involved in developing an X-ray laser (rapid spontaneous emission makes it hard to maintain a population inversion). It also explains why spectroscopy in the radio frequency (RF), microwave (MW) and, to a large degree, the infrared (IR) spectral regions is normally performed in absorption.

SPECTRAL LINESHAPES

The Bohr condition, Eq. (1), suggests the need for an exact match between the excitation frequency and the energy difference $E_2 - E_1$ but, in practice, a range of frequencies will be effective in producing a spectral transition. This frequency range defines the *linewidth* (or *absorption width*) and is affected by several contributions, the relative importances of which depend on the atom or molecule involved, the transition frequency and the experimental conditions. Amongst the main contributors are the natural linewidth, Doppler broadening and collisional broadening.

For an isolated atom, the natural width of a transition between two levels, one of which is the ground state, is determined solely by the radiative lifetime of the excited state. The energy spread is related to the uncertainty in the lifetime via the Uncertainty Principle:

$$\Delta E\, \Delta t = \hbar \quad , \tag{5}$$

and thus, via the Bohr relationship and assuming $\Delta t \approx \tau$,

$$\Delta v \approx (2\pi\tau)^{-1} \quad . \tag{6}$$

In the more general case of a transition between two excited levels, each of which has a finite natural lifetime, the overall linewidth is given by:

$$\Delta v \approx \frac{1}{2\pi}\left(\frac{1}{\tau_1} + \frac{1}{\tau_2}\right) \quad . \tag{7}$$

These expressions represent the natural full width half maximum (fwhm) of a spectral line. Natural line broadening makes a negligible contribution under most experimental conditions. Molecular transitions involving predissociated excited states provide one notable exception to this generalisation.

Collisions have the effect of reducing the lifetimes of excited states (quenching) and can thus make a contribution to the measured spectral linewidth. In the gas phase this is termed *pressure broadening*. For a single atom (A) the mean time between collisions with surrounding bath atoms (or molecules) B is given by Z^{-1}, where Z is the collision frequency. From the collision theory of gases,

$$Z = \sqrt{2} \, \sigma \bar{c} \, N_B \quad , \qquad (8)$$

where N_B is the number density of the collision partner B, σ is the collision cross-section and $\bar{c}$ is the mean speed. Taking representative values for σ (0.1 nm^2) and $\bar{c}$ (500 m s^{-1}) we obtain a value for $Z = 1.7 \times 10^9$ s^{-1} at 300 K and a pressure of 1 atmos. (101325 Pa). If we make the assumption that all collisions will cause a change in the quantum state of the atom (or molecule), then we arrive at a value of 9×10^{-3} cm^{-1} for the pressure broadening contribution to the spectral linewidth at this pressure. *Power* (or *saturation*) *broadening* is another possible source of spectral line broadening. This can occur in the presence of intense radiation fields since, under such conditions (if the frequency is appropriate), the atom will cycle rapidly between the two energy states connected by the transition. Thus, power broadening should always be considered as a possible problem when using lasers, or strong microwave or radio frequency sources.

These all constitute examples of *homogeneous* broadening mechanisms, so called because each atom in the sample can absorb radiation over the entire linewidth, *i.e.* the absorption and emission profile of the macroscopic assembly of atoms is the same as that of each of the individual atoms. Each gives rise to a *Lorentzian* spectral lineshape.

Doppler broadening is different in that its extent depends on the motion of the particular species of interest relative to the radiation field; it is thus termed an *inhomogeneous* broadening mechanism. For an individual atom (or molecule) moving towards the exciting radiation source the Doppler shifted absorption frequency is given by

$$\nu = \nu_0 \left[1 + \frac{v_z}{c} \right] \quad , \qquad (9)$$

where ν_0 is the absorption frequency of the stationary atom and v_z is the component of the molecular velocity along the light propagation axis (z). The quantity $\nu_0 v_z/c$ is referred to as the Doppler shift. For a molecule moving away from the radiation source the Doppler shift in Eq. (9) is negative. For molecules moving perpendicular to the light propagation axis there is no shift. A Maxwell-Boltzmann distribution of molecular velocities yields a *Gaussian* lineshape function, with a Doppler linewidth (fwhm, units s^{-1}) given by

$$\Delta\nu_D = \frac{2\nu_0}{c} \left[\frac{2RT\ln 2}{M} \right]^{1/2} \quad , \qquad (10)$$

where M is the molar mass (in kg). Note that $\Delta\nu_D$ increases linearly with frequency ν_0 and is thus most significant for high frequency transitions in light molecules at high temperatures.

Thus the overall lineshape exhibited by any particular spectral transition will contain contributions from a number of sources. Molecular spectra, particularly those involving heavier molecules, are often further complicated by the fact that many such transitions may overlap in frequency, leading to a broad, irregularly shaped and imperfectly resolved absorption feature. Such broad absorptions are the norm in spectra recorded in the condensed phase, but Doppler broadening is often the single most important factor limiting the resolution of spectra recorded in the gas phase. Some ways of alleviating this difficulty are illustrated amongst the examples considered later.

MEASURES OF ABSORPTION STRENGTH

Thus far we have concentrated on the frequencies and the lineshapes of atomic and molecular transitions. Now we consider what determines their intensities. Specifically, we develop the connection between the observable (*absorption coefficient*) and the calculable (*dipole-*, or *oscillator- strength*) measures of the transition strength.

Eq. (2) provides a convenient starting point, since B_{12} is directly proportional to the intrinsic absorptivity of the atom or molecule undergoing excitation [Eq. (4)]. Following Atkins (1983), the differential loss of intensity of radiation of frequency v due to an absorbing species with transition energy hv is directly proportional to the number density of species $N_1(v)$ in the initial state $|1\rangle$ that are capable of absorbing radiation at the frequency v, the radiation density, $\rho(v)$, at this frequency; it is also proportional to the differential pathlength $\delta\ell$:

$$-\delta I(v) = B_{12}N_1(v)\rho(v)hv\delta\ell \ . \tag{11}$$

This equation can be related to an operational quantity, the observed loss of intensity due to absorption by a sample of the atoms or molecules of interest:

$$-\delta I = \kappa(v)IC\delta\ell \ , \tag{12}$$

where $\kappa(v)$ is the molar Napierian absorption coefficient (units $m^2\ mol^{-1}$) and C the concentration. Eq. (12) is one version of the *Beer-Lambert law*, the integrated form of which may be found written as

$$\frac{I}{I_0} = \exp(-\kappa C\ell) \ , \tag{13}$$

or, in terms of the molar (decadic) absorption coefficient (often called the molar extinction coefficient), ε:

$$\frac{I}{I_0} = 10^{-\varepsilon C\ell} \ . \tag{14}$$

Eqs. (11) and (12) can be compared if we first note that ρ and I differ just by a factor of c, the speed of light. Thus Eq. (11) becomes[1] :

---

[1] Eqs. (13) and (14) hint at the possible confusion that can arise when discussing absorption coefficients. One frequently finds the symbol α used in place of κ in Eq. (13) but the IUPAC recommendation is that the symbol α should represent the (linear) Napierian absorption coefficient (units m^{-1}) - equivalent to the product κC. Further difficulties can arise because the path length ℓ has traditionally been measured in cm, and thus κ and ε generally appear in the older literature with units of $cm^{-1}\ dm^3\ mol^{-1}$. Finally, in many gas phase absorption measurements the 'concentration' is measured in pressure units (e.g. Torr or atmospheres), under which circumstances it is still common to see the proportionality constant in Eq. (13) quoted as k (with units of $pressure^{-1}\ cm^{-1}$), and with the temperature to which the pressure refers specified.

$$-\delta I(\nu) = B_{12}N_1(\nu)\frac{I(\nu)}{c}h\nu\delta\ell \ ,$$

which, if compared with Eq. (12), yields

$$\kappa(\nu)C = B_{12}N_1(\nu)\frac{h\nu}{c} \ .$$

When concerned with an incompletely resolved molecular absorption band it is customary to multiply both sides by $d\nu$ and then integrate over all frequencies of the absorption band to yield an expression:

$$\int_{band}\frac{\kappa(\nu)d\nu}{\nu} = \frac{B_{12}h\aleph}{cC} = \frac{B_{12}hL}{c} \ ,$$

where we have made use of the fact that the *total* number of absorbing molecules $\aleph$ is simply CL, where L is Avogadro's number. Finally, we recognise that for most absorption bands the frequency is essentially constant over the range for which $\kappa(\nu)$ is non-zero and we thus replace ν in the integral by a constant (ν_{12}, the centre frequency) to obtain the following expression for the integrated absorption coefficient A (units m^2 mol^{-1} s^{-1}):

$$A = \int_{band}\kappa(\nu)d\nu = B_{12}L\frac{h\nu_{12}}{c} \ . \tag{15}$$

Recalling Eq. (4), it is clear that we now have a link between an observable (A) and a quantity we can calculate, the transition moment $\langle 2|\mu|1\rangle$.

METHODS OF ABSORPTION SPECTROSCOPY

Here we summarise some of the various laser-based techniques for measuring absorption spectra. These will be discussed roughly in order of increasing frequency; thus techniques which are used to record infrared spectra will be introduced before methods which find more widespread use in the visible and ultraviolet spectral regions. Before embarking on this section it is perhaps worth emphasising that it is still the case that a large proportion (probably the major fraction) of all available atomic and molecular spectroscopic data has been obtained using non-laser methods (e.g. conventional spectrographs), and that Fourier transform infrared (FTIR) spectroscopy remains the premier means of obtaining a 'global' view of any particular infrared absorption spectrum.

(i) Direct Absorption Spectroscopy using Tunable Laser Sources

The basic absorption experiment (Fig.1) involves a radiation source (normally tunable and as close to monochromatic as possible), a sample cell (often a multipass, or 'White', cell to give a longer pathlength, and in some cases the sample is in the form of a molecular beam so as to give a colder, less rotationally congested spectrum) and a detection system.

Diode laser spectroscopy is one such technique that finds widespread use (Brown and Howard, 1992) in obtaining very high resolution infrared spectra of gas phase species. Fig. 2 shows part of the diode laser absorption spectrum of CS radicals generated by ArF laser photolysis of CS_2. This study serves to illustrate the sensitivity of the technique, since

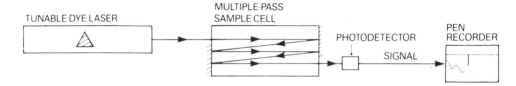

Figure 1. Schematic diagram of a laser absorption experiment (Andrews, 1990).

spectral analysis shows these photofragments to be formed in a very wide spread of internal energy states ($v \leq 11$, $J \leq 120$).

High resolution spectra of infrared transitions that occur at higher frequencies (for example C–H, O–H and N–H stretching fundamentals) can be measured using a *colour centre laser* or a *difference frequency laser spectrometer.* Difference frequency radiation in the wavelength range 2.2 - 4.2 µm can be generated by mixing the outputs of a single mode Ar^+ laser and a cw dye laser in a non-linear crystal (e.g. $LiNbO_3$). Figs. 3 and 4 show a typical difference frequency spectrometer set-up and a representative spectrum – part of the jet-cooled absorption spectrum of the v_1 (acetylenic C–H stretch) fundamental of 1-butyne. The overall appearance of the spectrum is very simple. This is not surprising, given the very low rotational temperature (*ca.* 6 K) of the sample. However, the very high experimental resolution ($\Delta \bar{v} \sim 0.0002$ cm^{-1}) enables one to see that each of the expected P and R branch features is in fact a many-line clump; this additional complexity is attributed to *intramolecular vibrational redistribution (IVR)* – i.e. to couplings between the 'bright' C–H stretching vibration and many of the levels in the 'dark' background of overtone and combination vibrational states.

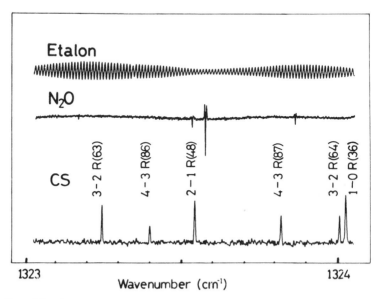

Figure 2. Part of the observed spectrum of CS radicals formed in the 193 nm photolysis of CS_2 (bottom trace). Individual peaks are labelled v'-v'' R(J''). The middle trace shows a spectrum of N_2O used as a frequency standard, whilst the top trace shows fringes from an etalon with free spectral range ~0.01cm^{-1} (Kanamori and Hirota, 1987).

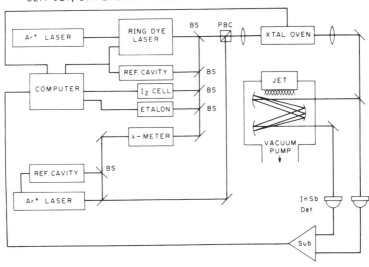

Figure 3. Schematic diagram of a difference frequency laser spectrometer designed to measure high resolution infrared spectra of jet-cooled molecules. BS indicates a beamsplitter, PBC a polarisation beam combiner (McIlroy and Nesbitt, 1989).

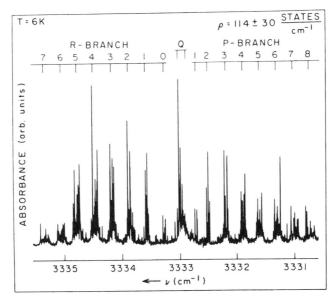

Figure 4. Acetylenic C–H stretch, $v = 1 \leftarrow 0$, spectrum of 1-butyne (McIlroy and Nesbitt, 1990).

Narrow bandwidth, tunable radiation throughout the visible and near ultraviolet spectral regions is available routinely from present day dye lasers; these are beginning to be supplemented by Ti-sapphire lasers and optical parametric oscillator sources. Using such lasers to record direct absorption spectra, in which one endeavours to measure a small diminution (ΔI) in the incident intensity (I), is perfectly feasible, but such is rarely done because it is generally possible to achieve higher sensitivity by recording some other property occurring as a result of the light absorption. Several such examples are considered in (iii) below.

Before leaving this section, however, we should introduce the idea of making absorption measurements *intracavity* (Fig. 5) —such techniques have found quite widespread use as a means of enhancing sensitivity to very weak absorptions, both in the infrared and the visible. Several factors contribute to this enhancement. One, of course, is simply the increase in the effective absorption path length as the radiation propagates back and forth within the cavity. Additionally, however, if the laser is arranged to be operating close to threshold then even a weak absorption (which can be viewed as an additional small loss process within the cavity) can have a dramatic effect on the overall gain and thus the output intensity.

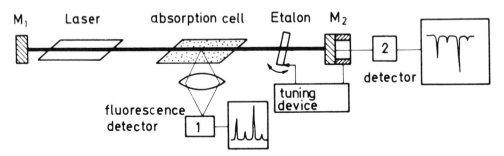

Figure 5. Intracavity absorption technique (Demtröder, 1991).

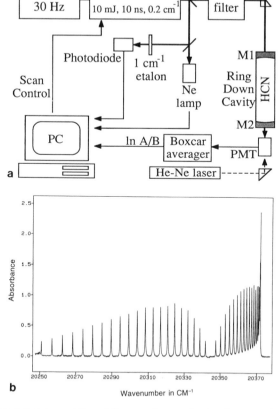

Figure 6. (a) Experimental arrangement used for recording cavity ring-down spectra of high overtone and combination bands of HCN. (b) Cavity ring-down spectrum of the 106 overtone of HCN (Romanini and Lehmann, 1993).

As a result, intracavity absorption spectroscopy can be an extremely sensitive probe of weak absorptions; its main disadvantage is the fact that the mechanisms for the sensitivity enhancement are too complex to allow any simple relationship between the intensity of the laser output and the sample absorbance. The same multipass advantage can be achieved using a passive resonator external to the laser cavity. Such is the basis of *cavity ring-down spectroscopy*, in which very weak absorption spectra (sensitivity *ca.* 10^{-9} cm^{-1}) can be obtained using a pulsed laser source by measuring the photon decay time within a high quality optical cavity as a function of excitation frequency. Fig. 6a shows a representative experimental arrangement. The decay time is inversely proportional to the resonator losses which, in a well designed cavity, will primarily be due to the (weak) absorption by the sample of interest. By way of illustration, Fig. 6b shows a room temperature spectrum of the (106) overtone band of HCN (*i.e.* absorption from the ground state to the combination level involving one C≡N stretching quantum and six quanta of C–H stretch) taken using the cavity ring-down technique.

(ii) Infrared Spectroscopy using Fixed Frequency Lasers

For completeness, at this stage we mention two atypical absorption techniques – *laser magnetic resonance (l.m.r.) spectroscopy* and *laser Stark spectroscopy* – both of which employ *fixed* frequency lasers and 'tune' the molecular transitions into resonance with this fixed frequency by application of a suitable external field. Both were pioneered at the time prior to the development of routinely tunable infrared laser sources, but both continue to find application today, especially in the far infrared.

The former relies on the Zeeman effect, and is generally applicable to molecules which possess a permanent magnetic dipole moment. A molecular energy level E_0 with total angular momentum J will split into (2J+1) components in an external magnetic field, B. The sub-level with magnetic quantum number M shifts from the zero field energy, E_0, by an amount

$$\Delta E = g\mu_B B M \ , \tag{16}$$

where μ_B is the Bohr magneton (9.2737 $\times$ 10^{-24} J T^{-1}) and g is the Landé factor. The frequency ν of the (v',J',M') ← (v",J",M") transition is therefore tuned by a magnetic field from its unperturbed frequency ν_0 to:

$$\nu = \nu_0 + \mu_B(g'M' - g''M'')B/h \ . \tag{17}$$

and we obtain three groups of lines (corresponding to ΔM = M'–M" = 0, ±1) associated with the (v',J',M') ← (v",J",M") transition. The achievable tuning range depends on B (obviously) and on the magnitude of (g'–g"); it is generally largest for molecules with a large permanent magnetic dipole moment. The largest magnetic moments are associated with electron spin; thus l.m.r. spectroscopy has found greatest use as a means of probing atoms and radical species possessing one or more unpaired electrons.

Fig. 7 shows a representative l.m.r. spectrometer, in which a cw CO_2 laser is used to pump a far infrared laser. The sample of interest is placed within the cavity of this latter laser (the lasing medium and the gaseous sample are kept apart by a thin polyethylene beamsplitter which also serves to polarise the laser radiation and the transitions) and the laser output is monitored as a function of the applied magnetic field. The sensitivity of this intracavity technique can be further enhanced, if necessary, by modulating the magnetic field and using phase sensitive detection. Fig. 8, which shows a portion of the far infra-red l.m.r. spectrum of the SH radical, illustrates the resolution and signal to noise ratios that can be achieved.

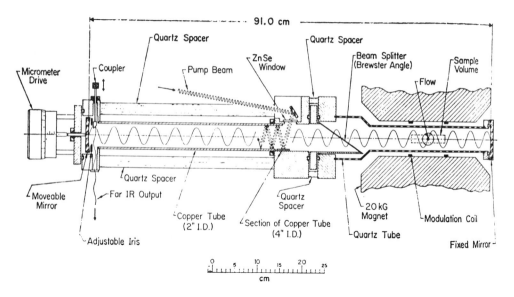

Figure 7. Laser magnetic resonance spectrometer, 40 - 1000 µm (Evenson, 1981).

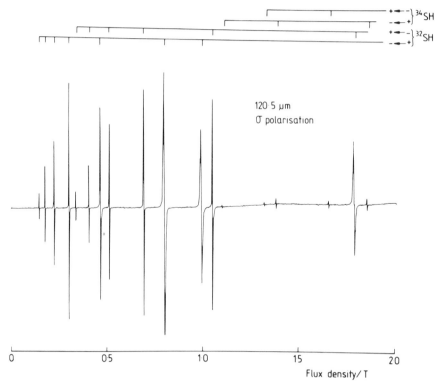

Figure 8. Far infrared l.m.r. spectrum of the SH radical, recorded over 0 - 2 Tesla with the 120.5 µm laser line of CD_2F_2 in σ polarisation ($\Delta M = \pm 1$). The rotational transition involved is $J = \frac{9}{2} \leftarrow \frac{7}{2}$ in the $^2\Pi_{3/2}$ component (Ashworth and Brown, 1992).

Laser Stark spectroscopy is the electric field analogue of l.m.r. spectroscopy. It is applicable to all polar molecules, *i.e.* molecules with a permanent electric dipole moment.

The extent to which a transition may be tuned depends on the magnitude of μ and on the strength of the applied electric field. By way of illustration, transitions of a molecule having a dipole moment of 2 Debye will tune by *ca.* 0.3 cm^{-1} in a field of 10^4 V cm^{-1}. Thus laser Stark spectroscopy is normally performed extracavity, in a cell equipped with two very closely spaced parallel electrodes (plate spacing as small as 1 mm). Fig. 9 shows an illustrative laser Stark spectrum, of part of the ν_4 band of $CH_3C^{15}N$, obtained using the P(25) CO_2 laser line near 906 cm^{-1} and applied electric fields in the range 0 – 15 kV cm^{-1}.

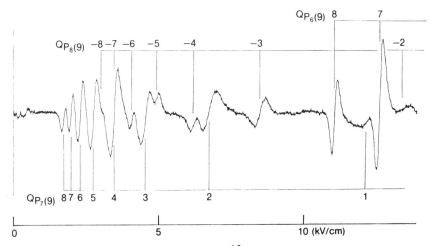

Figure 9. Laser Stark spectrum of the ν_4 band of $CH_3C^{15}N$, obtained using the P(25) CO_2 laser line near 906 cm^{-1}. $\Delta M = 0$ transitions of $^QP_6(9)$, $^QP_7(9)$ and $^QP_8(9)$ are observed (Mito *et al.*, 1984).

(iii) Specialised Absorption Techniques

This section considers some of the detection methods that rely on measuring the consequences of the absorption of radiation rather than the absorption itself. Laser induced fluorescence (LIF) is one such technique, but it is now available in so many variants that it is considered in its own sub-section below. Resonance enhanced multiphoton ionisation (REMPI) spectroscopy is another, which is considered fully in the Chapter by Ledingham.

Absorption introduces energy into individual molecules within a gas or liquid sample. Collisions cause energy transfer from the absorbing species into the bulk, which results in a local heating and thus a local increase in pressure. This pressure wave can be detected using a microphone. This is the basis of optoacoustic (or photoacoustic) spectroscopy. Fig. 10 shows a block diagram of a typical apparatus for recording an optoacoustic spectrum using a tunable cw laser. To enhance the signal to noise ratio the excitation is modulated using a chopper, and the spectrum is obtained by measuring the acoustic signal (with a phase sensitive detection system locked to the chopping frequency) as a function of excitation wavelength. Optoacoustic spectroscopy can also be performed with a pulsed laser; by way of illustration, Fig. 11 shows the optoacoustic spectrum of the weak fourth C –H stretching overtone transition of C_2H_2. Optoacoustic spectroscopy is applicable to gases (an inert buffer gas is often added to enhance energy transfer and the subsequent thermally induced pressure wave) and liquids. Organic solvents are generally preferable to water, since the high heat capacity of the latter tends to damp the acoustic signal.

Thermal lensing spectroscopy also relies on the local heating that results from photon absorption. The extent of heating scales with the incident laser intensity; thus the small volume of sample (normally a liquid) traversed by the most intense portion of the laser be-

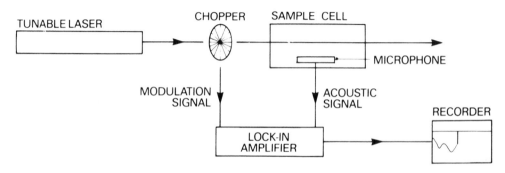

Figure 10. Block diagram of apparatus used for measuring the optoacoustic spectrum of a gas (Andrews, 1990).

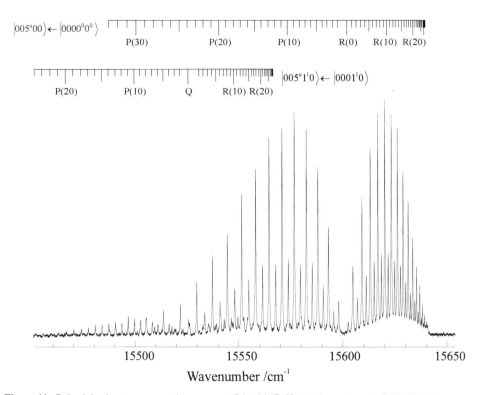

Figure 11. Pulsed dye laser optoacoustic spectrum of the 4th C–H stretch overtone in C_2H_2 (Wheeler, 1994).

am (the centre in the case of a TEM_{00} beam profile) experiences a greater temperature rise than sample at the periphery of the beam. These temperature gradients cause concomitant inhomogeneities in the local refractive index, which manifest themselves by causing the laser beam to defocus ('bloom'). Fig. 12 depicts the kind of experimental arrangement used for thermal lensing spectroscopy: an absorption spectrum is obtained by measuring the fraction of the incident beam that passes through an on-axis pinhole as a function of excitation wavelength.

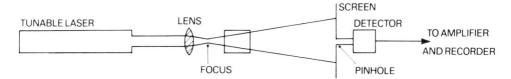

Figure 12. Schematic of apparatus used in thermal lensing spectroscopy. The 'blooming' that arises as the laser is tuned through an absorption band of the sample causes a spreading of the beam diameter which is observed as a decrease in the intensity passing through the pinhole (Andrews, 1990).

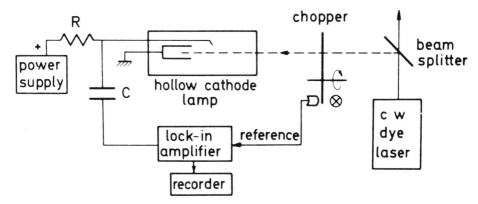

Figure 13. Experimental arrangement for optogalvanic spectroscopy in a hollow cathode lamp (Demtröder, 1991).

The final method to be considered in this section is *optogalvanic spectroscopy*. This relies on the change in electrical properties of a gas discharge, plasma or flame that can arise if excited with light of an appropriate frequency. Consider again two states $|1\rangle$ and $|2\rangle$ of an atomic species in a discharge. When a laser is tuned to resonance with the $|2\rangle \leftarrow |1\rangle$ transition it will cause a net transfer of population between the two states. If the two states have different ionisation probabilities then this population change will result in a change in the discharge current (hence the name) which can be detected as a voltage change across a ballast resistor (see Fig. 13).

The observed signal is sensitively dependent both on the changes in population of levels $|1\rangle$ and $|2\rangle$ and on their relative ionisation probabilities. Clearly, the signals observed can be both positive and negative. Increasing the incident light intensity can actually cause a signal to diminish or even change sign! Clearly, therefore, one should be cautious about using optogalvanic spectroscopy for quantitative population measurements. However, the technique does find widespread use as a means of wavelength calibration in laser spectroscopy. As an example, Fig. 14 shows part of the rich optogalvanic spectrum of atomic neon.

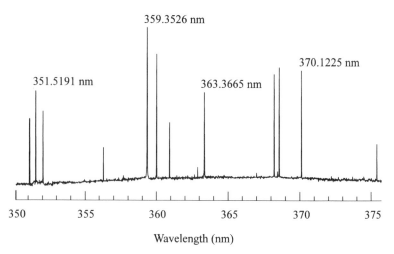

Figure 14. Part of the optogalvanic spectrum of atomic neon excited in a hollow cathode lamp (Morgan, 1994).

EMISSION SPECTROSCOPY; LASER INDUCED FLUORESCENCE

We have now considered a wide range of very sensitive laser spectroscopy techniques. However, *sensitivity* is only one of the needs of the analytical chemist. *Selectivity* is another and, in this regard, all of the absorption methods considered thus far are limited in that the only species-selective aspect to the measurement is the particular wavelength(s) at which the absorption is measured. Hence the appeal of *laser induced fluorescence (LIF)* – a technique which relies on measurement of the excited state fluorescence that results following photo-excitation (Pfab, 1992). The principle of the method is illustrated by the energy level diagram shown in Fig. 15a. An 'absorption' spectrum is obtained by measuring the excitation spectrum for creating fluorescing excited state molecules. This offers several advantages:

• *Quantitative Concentration Measurements.* Provided the exciting laser is of sufficiently low intensity that saturation is not a problem, the *total* LIF intensity (*i.e.* without resolution of the emission spectrum), given by

$$I_{LIF} \propto I_{laser} \, N_1 \, B_{12} \, \Phi \ , \tag{18}$$

(where B_{12} is the Einstein coefficient for absorption and Φ is the fluorescence quantum yield of the excited state $|2\rangle$), is directly proportional to the number density in the initial state undergoing excitation.

• *Sensitivity.* Unlike most forms of absorption spectroscopy it is, in principle at least, a zero background technique. If the atom or molecule of interest does not absorb at a particular wavelength then zero LIF should result. In practice, scattered laser light can be minimised (by using appropriate filters in front of the detector or, in the case of pulsed laser experiments, by gating the detection electronics so as to only collect signal after the laser excitation pulse) but rarely truly eliminated.

• *Enhanced selectivity.* If the emission step connects two discrete levels(*i.e. bound* levels in the case of a molecule) then, from Eq. (1), it will occur only at certain well-defined frequencies. Thus the technique is species-selective both in the excitation step *and* in the fluorescence (detection) step.

• *Spatial localisation.* Fluorescence can be imaged from a localised volume (*c.f.* absorption techniques where one is sensitive to species along the entire column length).

At this stage, it is also sensible to point out the principal limitation of the LIF technique – namely, the requirement that the excited state must have a significant fluorescence quantum yield, preferably in an operationally convenient wavelength range. This quantum yield is defined as the ratio of the number of fluorescent photons to the number of photons absorbed. Thus the technique is applicable to the detection of trace concentrations of most atomic species, for which Φ will generally be at least close to unity, but, in the case of molecules, it is only appropriate for those species that possess suitable fluorescing excited states. This requirement precludes use of LIF as a detection method for some small gas phase species (*e.g.* the alkyl radicals or methylene in its triplet ground state, for which all known excited states are predissociated) as well as for many large molecular species whose excited states are subject to fast non-radiative decay (*e.g.* internal conversion, intersystem crossing). It can also prove a limitation in high pressure situations, where collisional quenching will be the dominant excited state loss mechanism.

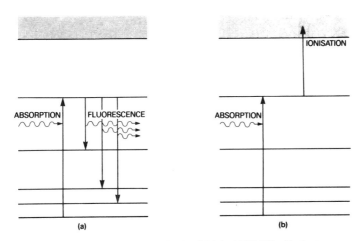

Figure 15. Energy level diagrams illustrating (a) LIF and (b) 1+1 REMPI . Excitation spectra are obtained by measuring (a) the total fluorescence or (b) the ion (or photoelectron) yield as a function of excitation wavelength. Extra information about the intervening energy levels can be derived from analysis of the wavelength dispersed emission in (a), whilst additional information about the energy levels of the ion can be obtained by analysing the kinetic energies of the photoelectrons formed in (b) (Andrews, 1990).

Fortunately, with the development of increasingly intense tunable lasers, many such species can now be detected using the complementary technique of *resonance enhanced multiphoton ionisation (REMPI) spectroscopy*. Fig. 15b shows the excitation scheme appropriate to the case of 1+1 REMPI. The REMPI technique can actually offer a number of advantages over LIF spectroscopy. Sensitivity is one. This stems from the fact that we detect ions (which can be collected with 100% efficiency) rather than photons, which can never be detected with anything like the same efficiency. The technique also offers good species selectivity. This arises because (a) the structure appearing in the spectrum of ion yield versus excitation wavelength arises as a result of the resonance enhancement associated with the first bound ← bound excitation step, and thus provides a characteristic spectral signature of the particular neutral molecule under study; (b) the mass of the parent ion can normally be deduced if the resulting ions are mass analysed (e.g. by time-of-flight mass spectrometry) and/or (c) the ionisation potential of the species under investigation can normally be measured (checked) if one chooses to monitor the resulting photoelectrons rather than the partner ions. Recent reviews by Ashfold et al. (1993, 1994) serve to illustrate some of the spectroscopic opportunities offered by the technique.

LIF can be used for *trace elemental analysis,* but here too it is progressively being superceded by REMPI. The sample to be analysed (often in solution) is atomised in a plasma, in a furnace or in a flame (laser ablation methods are finding increasing use as a means of atomising solid samples) and a spectrum of the atomic species then recorded using a tunable laser. Fig. 16 shows a representative experimental arrangement. The resulting spectrum can be quite complex given the range of atoms present, and the fact that the atomisation process will often lead to population of both ground and excited states, but the fluorescence frequencies normally provide an unambiguous fingerprint of the elements present. Both sensitivity and selectivity can be improved using two (or more) colour multiply resonance enhanced MPI; increasingly this is becoming the method of choice for this type of elemental analysis.

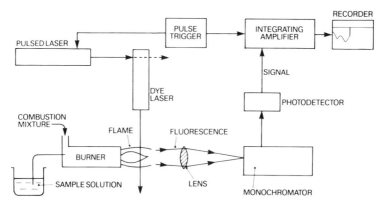

Figure 16. Schematic diagram of instrumentation required for elemental LIF measurements using a pulsed dye laser and flame atomisation of the sample solution (Andrews, 1990).

LIF still finds widespread use in the determination of spatial and/or temporal concentration profiles of atomic (and some simple molecular) species in, for example, plasmas, flames and discharges (see Fig. 17) – environments in which there may already be a significant concentration of ions. *Flame temperatures* can be deduced from measurements of the relative intensities of LIF transitions which start from different energy states of the same atom or, more commonly, by measuring the relative intensities of different rovibronic transitions in the LIF spectra of simple molecular species like OH, CH, NH, C_2 or CN. Converting the observed line intensities in the LIF excitation spectrum of a molecular system into relative populations of the various quantum states requires detailed knowledge of the relevant molecular spectroscopy – *e.g.* accurate *ro*vibronic transition probabilities (Franck-Condon factors, rotational line strengths *etc.*) – and a thorough understanding of any level-dependent quenching efficiencies, predissociation, *etc.* which might render Φ itself a level-dependent quantity. Given such information and using the Boltzmann equation:

$$\frac{N_i}{N_0} \propto \frac{g_i}{g_0} \exp\left[-(E_i - E_0)\Big/kT\right] , \tag{19}$$

(which relates the relative population of the state $|i\rangle$, with energy E_i and degeneracy g_i, to that of the ground state $|0\rangle$, and will hold for any species in thermal equilibrium) then, from Eq. (18), it follows that the temperature can be obtained from the slope of a plot of

$\ln[I_{LIF}/B\Phi]$ versus E_i. Fig. 18, which shows part of the LIF excitation spectrum of $NH(X)_{v=0}$ radicals in a C_2H_2/N_2O flame, illustrates this application.

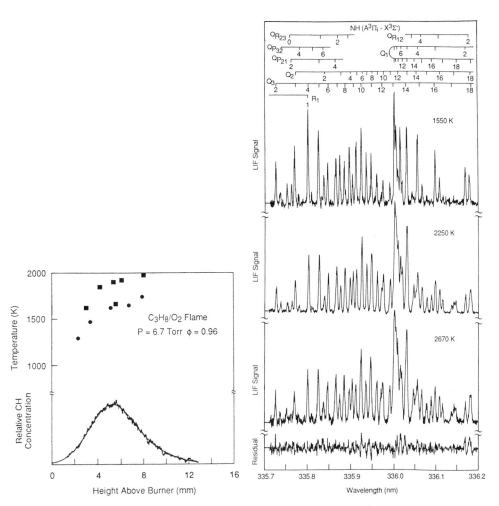

Figure 17. CH radical concentration profile for a 6.7 Torr propane/O_2 flame, demonstrating that the CH radical is present only in the reaction zone. The boxes and circles show temperatures determined from analysis of rotationally resolved LIF excitation spectra of, respectively, the $CH(A^2\Delta - X^2\Pi)$ and $CH(B^2 - X^2\Pi)$ transitions (Rensberger et al.,1989).

Figure 18. LIF excitation spectra of the Q branch region of the $NH(A^3\Pi - X^3)$ transition for a 13.8 Torr C_2H_2/N_2O flame. The spectra (top to bottom) were taken at 1.1 mm, 4.0 mm and 7.3 mm above the burner and are fitted to temperatures of 1550 K, 2250 K and 2670 K respectively. The bottom trace is the residual of the fit at 2670 K (Rensberger et al.,1989).

At this point we should mention that *coherent anti-Stokes Raman spectroscopy (CARS)* (see, for example, Berger et al. (1992)) is another well established laser based diagnostic for combustion processes, and that there is now increasing interest in the related technique of *degenerate four wave mixing (DFWM) spectroscopy* (see, for example, the overview by Farrow and Rakestraw (1992)). Neither of these latter spectroscopies show the ultimate sensitivity of LIF, but each has the advantage that the signal emerges as a coherent, and therefore directional, beam which can be highly advantageous when attempting to probe a bright, luminous sample like a flame or a plasma.

Laser induced fluorescence also finds use in the analysis of eluant from liquid chromatography and in laser fluorimetry (where the species of interest is first 'tagged' with a suitable fluorescent dye molecule). However, its greatest application to date has been in the field of gas phase chemical physics, particularly small molecule spectroscopy (e.g. in the determination of vibrational, rotational, spin-orbit constants etc., and of transition probabilities – Franck-Condon factors, radiative lifetimes, etc.), in studies of collisional energy transfer processes and in probing the products of unimolecular (e.g. photofragmentation) and bimolecular chemical reactions. We end this Chapter with some illustrative examples of such studies.

Fig. 18 showed part of the LIF excitation spectrum of the NH free radical following excitation on its $A^3\Pi - X^3\Sigma^-$ transition in the near ultraviolet. Numerous such molecular spectra have been recorded and analysed in great detail to provide a wealth of spectroscopic (and thus structural) data about the particular molecule under investigation. The widespread availability of supersonic molecular beam sources, and the rotational (and vibrational) cooling they provide, means that such studies are also worthwhile on much larger molecular species, including transients and ions. As the top trace in Fig. 19 shows, the density of rovibronic lines in the room temperature absorption spectrum of a heavy species like $C_6F_6^+$ is such that, even at Doppler limited resolution, the spectrum appears essentially continuous. Jet expansion, insofar as it depopulates many of the levels that contribute to the 'warm' spectrum, provides one way of simplifying such spectra. An alternative strategy, which does not involve removal of many of the spectroscopic transitions, is to improve the resolution. This circumvents the Doppler broadening, which tends to be the single most important factor limiting the resolution of gas phase electronic spectra.

There are a number of ways of alleviating Doppler broadening. One is to make use of the directionality (rather than, necessarily, the rotational cooling) provided by a skimmed molecular beam. Fig. 20 illustrates the state of the art resolution that can be achieved using a narrow bandwidth laser ($\Delta\nu \sim 1$ MHz), probing at right angles to a triply skimmed supersonic beam of ICl in Ar. Given such collimation, the spread in velocities along the laser propagation axis will necessarily be greatly reduced. The 5 MHz (1.7×10^{-4} cm^{-1}) spectral resolution achieved is sufficient to reveal not just the iodine hyperfine structure but also the much smaller chlorine hyperfine splittings. The alternative is to use one of the many forms of *sub-Doppler spectroscopy* that are now available.

Many variants of sub-Doppler laser spectroscopy involve use of an intense narrow band laser ($\Delta\nu_{laser} \ll \Delta\nu_D$) to selectively saturate a part of the velocity distribution of the absorbing molecules, with successive probing of this selective 'hole burning' by a monochromatic tunable probe wave. Such techniques are discussed at some length in Demtröder's classic text (Demtröder, 1991). Here we mention just one other sub-Doppler technique – two photon spectroscopy. It has the virtues that (*a*) unlike saturation methods, *all* molecules in the absorbing state, irrespective of their velocities, contribute to the Doppler free transition and (*b*) bulk samples may be used, provided that the pressure is maintained sufficiently low for collisional broadening to be unimportant. As with the saturation spectroscopies, the experiment employs counterpropagating beams of identical frequency (as measured in the laboratory frame). However, as seen by an absorbing molecule moving with a velocity v_z (z component), the frequency of one beam appears to be shifted by a factor $[1 + (v_z/c)]$ but that of the opposing beam shows a decrease by a factor of $[1 - (v_z/c)]$. Thus for two photon excitation brought about by absorption of *one photon from each of the two beams* the overall gain in energy ΔE is given by:

$$\Delta E = h\nu\left[\left(1+\frac{v_z}{c}\right)+\left(1-\frac{v_z}{c}\right)\right] = 2h\nu \ , \tag{20}$$

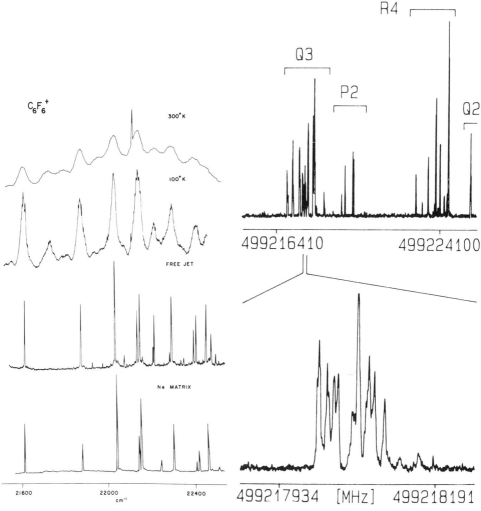

Figure 19. Laser induced fluorescence spectrum of the $C_6F_6^+$ ($\tilde{A}-\tilde{X}$) transition taken under a variety of conditions. From top to bottom: (i) $C_6F_6^+$ ions produced by Penning ionisation at room temperature, (ii) as (i) but with a flow system maintained at liquid nitrogen temperature, (iii) $C_6F_6^+$ ions pro-duced by two photon ionisation by an ArF laser in a free jet expansion, (iv) $C_6F_6^+$ ions in a Ne matrix at 4 K. The matrix spectrum has been shifted in absolute frequency so that its origin coincides with that of the gas phase spectra (Miller and Bondybey, 1983).

Figure 20. LIF excitation spectrum of a congested 0.3cm^{-1} portion of the ICl(A–X) (19,0) band with (below) an expanded view (0.01 cm^{-1}) of the hyperfine structure within the Q(3) line (Johnson et al., 1990).

independent of v_z (to first order in v_z/c). By way of illustration, Fig. 21 shows part of the 'Doppler-free' two photon spectrum of two Q branches in the $\tilde{A}-\tilde{X}$ system of C_6H_6.

Fig. 21 is also interesting because it emphasises the point that an LIF excitation spectrum is only equivalent to an absorption spectrum *if* the fluorescence quantum yield is the same for all the excited levels. Clearly, there are far fewer lines in the lower spectrum. Analysis shows all lines in the $14_0^1 1_0^2$ band (Fig. 21b) to involve upper state levels with K' = 0, from which we deduce that at this level of internal excitation ($E_{vib} \sim 3412$ cm^{-1}), levels with K' > 0 decay non-radiatively on a timescale that is fast compared with the fluorescence lifetime. Yet the spectrum of the $14_0^1 1_0^1$ band (Fig. 21a) showed no such selectivity. This is just one illustration of the way in which molecular rotation (and the Coriolis coupling it promotes) can affect the stability of molecular excited states. Fig. 22 shows another example, involving the much studied HNO free radical. The LIF excitation spectrum shown (Fig. 22b) shows a clear 'breaking off' of the rotational structure above K' = 4, J' = 11 in the (100) vibrational level of the $\tilde{A}$ state, even though this structure persists to higher

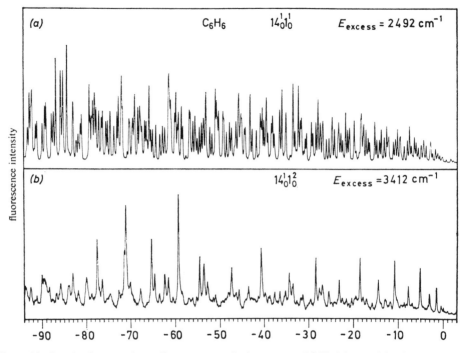

Figure 21. Doppler-free two photon fluorescence excitation spectra of C_6H_6: (a) part of the Q branch of the ($\tilde{A}$–$\tilde{X}$) $14_0^1 1_0^1$ band (E_{vib} = 2492 cm^{-1}); (b) corresponding part of the $14_0^1 1_0^2$ band (E_{vib} = 3412 cm^{-1}) (Riedle et al., 1983).

J' and K' values in the absorption spectrum (Fig. 22a). This reflects the onset of *predissociation* in this molecule; the analysis of the breaking off limits in several such K'–K" sub-bands has allowed accurate determination of D_0(H–NO) = 16450 ± 10 cm^{-1}.

Thus far our examples have concentrated on the kinds of information that can be gleaned from measuring the LIF excitation spectrum. Fig. 15a hinted at the additional information that can be obtained if we take the trouble to wavelength resolve the LIF. Fig. 23, which shows the wavelength resolved emission spectrum of Na_2 molecules following Kr$^+$ laser excitation to the v' = 34, J' = 50 level of the $A^1\Sigma_u^+$ state, provides an illustrative example. Analysis yields term values for a very wide spread of ground state levels, (spanning v" = 4 right up to the dissociation limit), from which it is possible not just to obtain an accurate value for D_0(Na–Na), but also to derive a very clear picture of the ground state potential energy function.

Dispersed emission spectroscopy has a number of limitations. Clearly, the technique is restricted to the study of levels lying at lower energy, and to which there is significant spontaneous transition probability. Further, the resolution of such spectra will be limited (due to the intrinsic weakness of any emission, and the low photon collection efficiency). Two colour double resonance spectroscopies (Fig. 24) can provide a route to circumventing all of these difficulties. The particular form of double resonance spectroscopy that has closest analogy with dispersed emission is *stimulated emission pumping (SEP)* (Fig. 24c). When the probe (or 'dump') laser is on resonance, it stimulates emission in the direction of

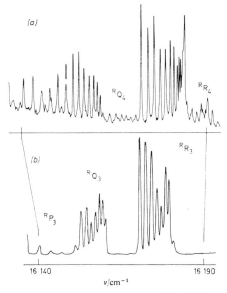

Figure 22. The (100) - (000) K' = 4 ← K" = 3 sub-band of the $\tilde{A}^1A"$– $\tilde{X}^1A'$ band system of HNO recorded (*a*) in absorption and (*b*) as an LIF excitation spectrum. The weak $^rQ_4$ and $^rR_4$ branches in (*a*) belong to the (020) - (000) K' = 5 ← K" = 4 sub-band (Dixon et al., 1981).

the probe laser beam, thereby depleting spontaneous fluorescence. The resonance condition can thus be found via detection of the fluorescence side-light without the losses inherent in the use of a monochromator. Alternatively, if the pump laser is sufficiently intense to cause REMPI, the SEP transitions may be observed as dips in the ion yield as the dump laser frequency is scanned. The achievable resolution is much higher than in dispersed LIF – being ultimately limited by the bandwidths of the lasers employed in most cases. Fig. 25 shows two SEP spectra of formaldehyde. Analysis of many such spectra has provided a wealth of information about the pattern of highly excited rovibrational energy levels in the ground electronic states of many small polyatomic molecules, *e.g.* H_2CO (Fig. 25), C_2H_2, HCN, and radicals, e.g. C_3, NCO (see Northrup and Sears (1992)). In passing, we mention the considerable recent interest in a two colour variant of four wave mixing spectroscopy known as *two colour laser induced grating spectroscopy (TC-LIGS)* which, in favourable cases, can be a high signal to noise rival to SEP.

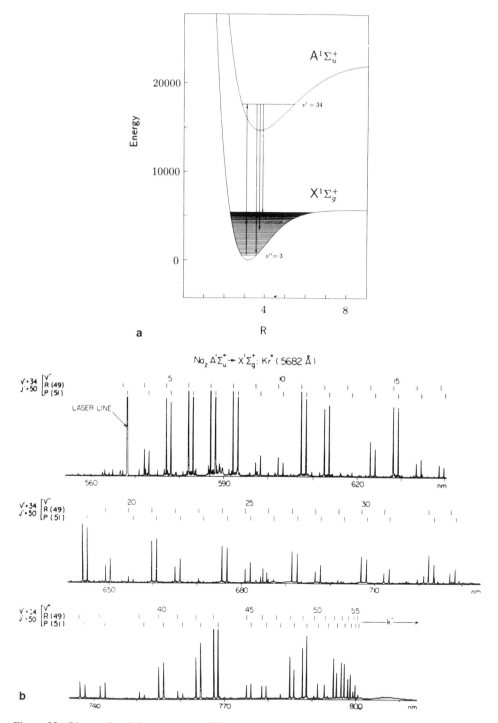

Figure 23. Dispersed emission spectrum of Na$_2$ vapour following excitation with the 568.2 nm line of a Kr$^+$ laser. This wavelength excites Na$_2$ molecules from the v" = 3, J" = 51 level of the $X^1\Sigma_g^+$ ground state to the v' = 34, J' = 50 level of the $A^1\Sigma_u^+$ excited electronic state, as shown in (a). The shown fluorescence lines (b) arise as a result of transitions from this level to all bound ground state levels with v" ≥ 4 (one P and R line to each v" level); the spectrum also shows a broad, weak feature in the near infrared associated with emission to continuum levels above the dissociation limit (Verma et al., 1983, and Struve, 1989).

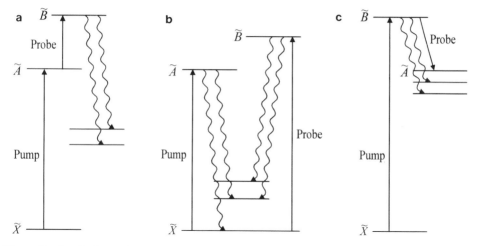

Figure 24. Level schemes for two colour optical-optical double resonance (OODR): either the pump laser or the probe laser may be scanned in wavelength whilst the frequency of the other is held fixed. (*a*) sequential OODR, (*b*) competitive excitation, (*c*) stimulated emission pumping.

As Fig. 24 suggests, SEP is just one of several variants of OODR spectroscopy. OODR techniques also find widespread use in simplifying congested spectra involving highly excited states, and as a means of circumventing Franck-Condon restrictions. Fig. 26 provides one somewhat unusual example. Individual rotational states associated with the $v' = 3$ level of

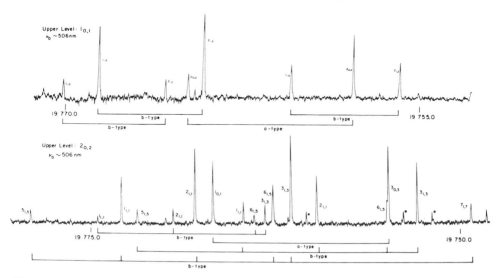

Figure 25. SEP spectra (I_{fluor} decreases as one scales the vertical axis) of formaldehyde at ground state energies near 8530 cm^{-1}. The spectra from the $J'_{K'_a K'_c} = 1_{0,1}$ and $2_{0,2}$ intermediate levels enable unambiguous assignment to be made of three *b*-type and one *a*-type bands (Hamilton et al., 1986).

57

the E state of NO are predissociated by interaction with a near resonant quasi-bound state which acts as a doorway into the full continuum of states associated with the products N + O. The predissociation efficiency depends on the 'detuning' of the interacting levels which, as Fig. 26 illustrates, results in a rovibrational level dependence to the rate of predissociation of the E state of NO.

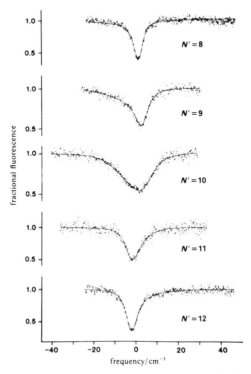

Figure 26. OODR fluorescence dips observed for transitions in the NO $E^2\Sigma^+ - X^2\Pi$ (3,3) band involving upper state levels with N' = 8 - 12. The solid curves are least-squares fits to the (asymmetric) experimental line profiles (Ashfold et al., 1986).

VECTORIAL CONSEQUENCES OF PHOTON ABSORPTION

The foregoing has concentrated largely on molecular spectroscopy *per se,* but there are numerous instances where molecular spectra are recorded, either by LIF or REMPI, not for their own sake but in order to probe dynamical aspects of chemical change, *e.g.* desorption of a gas molecule from a surface, molecular photochemistry, collisional energy transfer processes, bimolecular reactions. These applications rely on the fact that photo-excitation, the probability of which depends on the square of the dot product of μ and ε, is by definition an anisotropic process. This anisotropy can manifest itself in many ways: indeed, the study of vectorial (rather than scalar) consequences of photo-excitation is a 'hot' topic in contemporary chemical physics.

We begin with a simple example. Though one might start with an isotropic sample of molecules in (say) a gas, irradiation with linearly polarised laser light will cause preferential excitation of those molecules whose transition dipole happen to be pointing parallel to the ε

vector of the light. The initial distribution of excited molecules will thus be anisotropic. This anisotropy (and, for example, its decay through collisions) can be studied by measuring the degree of polarisation of the excited state fluorescence.

Now consider photo-excitation to an excited state that is dissociative. For any parent molecule there is a fixed angular correlation between the direction of the transition dipole moment and the direction of fragment recoil; usually the recoil is along the bond that breaks during the dissociation process. If the bond breaks on a timescale that is fast compared with the rotational period of the parent, then the alignment of μ in the laboratory frame (because of the $\mu.\varepsilon$ dependence of the photo-excitation process) should lead to an alignment of the fragment recoil velocity vector, $\mathbf{v}$. The angular distribution of these velocities is traditionally written

$$I(\theta) = \frac{1}{4\pi}\left[1+\beta P_2(\cos\theta)\right] , \qquad (21)$$

where θ is the angle between $\mathbf{v}$ and ε, $P_2(\cos\theta) = \frac{1}{2}(3\cos^2\theta-1)$ is the second Legendre polynomial and β is a parameter that describes the degree of anisotropy which takes limiting values of +2 (for $\mathbf{v}$ parallel to ε) and -1 ($\mathbf{v}$ perpendicular to ε). Figure 27 illustrates, schematically, how this *recoil anisotropy* will affect the Doppler profiles of individual features in the spectrum of a photofragment, whilst Fig. 28 shows a 'real-life' example − an

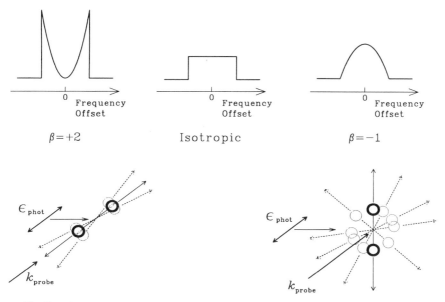

Figure 27. Doppler profiles predicted for photofragments recoiling with a single speed, v, but different degrees of spatial anisotropy characterised by, respectively, β = +2, 0 and -1. This example assumes an orthogonal photolysis - probe beam geometry (i.e. $\mathbf{k}_{probe} \perp \mathbf{k}_{phot}$) and that the ε vector of the photolysis laser lies in the plane of propagation of the two laser beams. Under these circumstances, the products of a fragmentation characterised by an anisotropy parameter, β, of +2 (consistent with dissociation following excitation via a *parallel* transition) will fly preferentially towards and away from the probe laser, leading to the symmetrically split, forward-backward peaking Doppler profile shown. Conversely, a dissociation characterised by β = 0 will yield an isotropic spatial distribution of recoiling fragments (and a flat-topped Doppler profile), whilst the fragments from a dissociation with β = -1 (i.e. where $\mathbf{v}$ is at right angles to μ, and thus ε_{phot}) will tend to recoil in a plane perpendicular to $\mathbf{k}_{probe}$ and thus give rise to a symmetrical Doppler profile peaked at the line centre frequency (Ashfold et al.,1992).

LIF spectrum of the Doppler lineshape of OH(X) fragments in their $\Omega = \frac{1}{2}$, v" = 0, N" = 5 level, resulting from the 266 nm photodissociation of gas phase H_2O_2 molecules, monitored via the $Q_2(5)$ line of the OH($A^2\Sigma^+ - X^2\Pi$) (0,0) transition. The Doppler splitting in this particular spectrum has been accentuated by use of a technique known as *velocity aligned Doppler spectroscopy (VADS)*. In Doppler spectroscopy one measures the spread of velocities along the propagation axis of the probe laser (recall Fig. 27). By introducing a time delay between the photolysis event and the probe laser pulse one discriminates against those fragments whose recoil velocity vectors do not point exactly parallel or antiparallel to k_{probe}. This is the basis of velocity aligned Doppler spectroscopy.

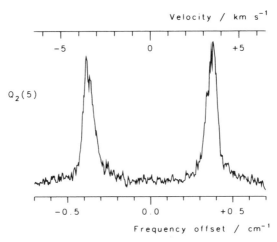

Figure 28. Doppler split LIF excitation lineshape showing the forward and reverse velocity component for OH(X) fragments resulting from pulsed 266 nm photolysis of H_2O_2 after a delay of 600 ns. The frequency zero is the absorption frequency of an OH radical at rest, 32338.86 cm^{-1} for the $Q_2(5)$ probe transition used here (Dixon et al., 1988).

A little thought should suffice to convince one that the *rotational* motion of a molecular photofragment (in a given J state) can have a similar anisotropic distribution in space. This anisotropy is usually characterised in terms of a rotational *alignment parameter*, $A_0^{(2)}$, which is the rotational analogue of β but has limits $\frac{2}{5}$ as great (*i.e.* $A_0^{(2)}$ takes values in the range $+\frac{4}{5}$ for $\mathbf{J} \parallel \boldsymbol{\mu}$ and $-\frac{2}{5}$ for $\mathbf{J} \perp \boldsymbol{\mu}$). Since the transition moment of a molecule is fixed in the molecular frame, and thus rotates with the molecule, we can probe the alignment of the fragment using polarised LIF excitation, varying the relative polarisations of the photolysis and probe lasers. To illustrate this point, consider the two LIF excitation spectra shown in Fig. 29. Both are spectra of the nascent OH(X) photofragments resulting from the 355 nm photolysis of gas phase nitrous acid molecules. Both were recorded using counterpropagating photolysis (p) and analysis (a) laser beams, with the resulting OH(A→X) LIF detected at right angles using a photomultiplier that accepts all fluorescence polarisations with equal efficiency. The only difference, as the insets show, is that the spectrum shown in Fig. 29a was recorded with ε_p and ε_a orthogonal to one another, with the fluorescence detected along an axis parallel to ε_a, whilst in Fig. 29b ε_p was rotated through 90° so as to be parallel to ε_a. As Fig. 29 shows, the former experimental geometry appears to favour the detection of P and R branch lines at the expense of Q branch transitions. Such variation in the relative intensities of P (or R) versus Q transitions in the LIF spectrum of the fragment is a classic signature of its alignment. Careful comparisons of these relative line intensities indicates that, for the OH(X) fragments arising in the near UV photolysis of *trans-* HONO, $A_0^{(2)}$ takes small positive values; i.e. there is a tendency for $\mathbf{J}_{OH}$ to be aligned parallel to $\boldsymbol{\varepsilon}_p$, and thus $\boldsymbol{\mu}_{parent}$. This is understandable. The HONO($\tilde{A} - \tilde{X}$) transition moment is known to lie perpendicular to the molecular plane. The forces acting during the dissociation will lie in the plane of the excited molecule so, in the limit of prompt dissociation, $\boldsymbol{\varepsilon}_p$ should be parallel to $\mathbf{J}_{OH}$ (in the classical, or high J, limit).

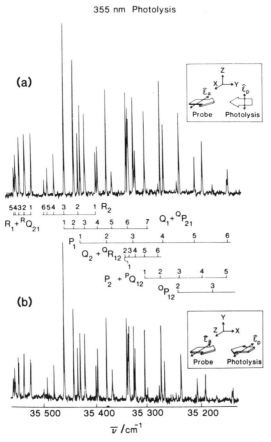

Figure 29. LIF excitation spectra of OH(X)$_{v=0}$ fragments resulting from HONO photolysis at 355 nm, with (*a*) $\boldsymbol{\varepsilon}_p \perp \boldsymbol{\varepsilon}_a$, and (*b*) $\boldsymbol{\varepsilon}_p \parallel \boldsymbol{\varepsilon}_a$. In both cases, the fluorescence is detected along an axis parallel to $\boldsymbol{\varepsilon}_a$ (Vasudev et al., 1984).

Like the recoil anisotropy, this vector correlation relies on having $\boldsymbol{\mu}$ aligned in the laboratory frame, and so the magnitude of $A_0^{(2)}$ degrades rapidly if the fragmentation does not occur promptly. We end by pointing out that, in many cases the story can get yet more complicated! Since we recognise the possible correlation between $\boldsymbol{\varepsilon}$, $\boldsymbol{\mu}$ and $\mathbf{v}$ (the recoil anisotropy) and between $\boldsymbol{\varepsilon}$, $\boldsymbol{\mu}$ and $\mathbf{J}$ (the alignment), it is clear that there could also be a correlation between $\mathbf{v}$ and $\mathbf{J}$. This is an especially interesting vector correlation since it depends only on the forces acting at the instant of dissociation and thus is not degraded in situations where the excited state (pre)dissociates over an extended time scale. This correlation is, in fact, the most intimate monitor of the act of molecular photofragmentation; in favourable cases it, too, can be quantified by careful measurement of the relative intensities and shapes of individual P (or R) and Q lines in the excitation spectrum of the fragment.

REFERENCES

Andrews, D.L., 1990, "Lasers in Chemistry," 2nd ed., Springer-Verlag, Berlin.

Ashfold, M.N.R., Dixon, R.N., Prince, J.D., Tutcher, B., and Western, C.M., 1986, *J. Chem. Soc. Faraday Trans. 2*, 82:125.

Ashfold, M.N.R., Lambert, I.R., Mordaunt, D.H., Morley, G.P., and Western, C.M., 1992, *J. Phys. Chem.* 96:2938.

Ashfold, M.N.R., Clement, S.G., Howe, J.D., and Western, C.M., 1993, *J. Chem. Soc. Faraday Trans.* 89:1153.

Ashfold, M.N.R., and Howe, J.D., 1994, *Ann. Rev. Phys. Chem.* 45:57.

Ashworth, S. and Brown, J.M., 1992, *J. Mol. Spect.* 153:41.

Atkins, P.W., 1983, "Molecular Quantum Mechanics," 2nd ed., Oxford University Press, Oxford.

Berger, H., Lavorel, B., and Millot, G., 1992, Nonlinear raman spectroscopy, *in*: "Applied Laser Spectroscopy: Techniques, Instrumentation and Applications," D.L. Andrews, ed., VCH Publishers, Inc., New York.

Demtröder, W., 1991, "Laser Spectroscopy", 2nd ed., Springer, Heidelberg.

Dixon, R.N., Noble, M., and Taylor, C.A., 1981, *Farad. Disc. Chem. Soc.* 71:125.

Dixon, R.N., Nightingale, J., Western, C.M., and Yang, X., 1988, *Chem. Phys. Lett.* 151:328.

Evenson, K.M., 1981, *Farad. Disc. Chem. Soc.* 71:7.

Farrow, R.L., and Rakestraw, D.J., 1992, *Science* 257:1894.

Hamilton, C.E., Kinsey, J.L., and Field, R.W., 1986, *Ann. Rev. Phys. Chem.* 37:493.

Howard, B.J., and Brown, J.M., 1992, High-resolution infrared spectroscopy, *in*: "Applied Laser Spectroscopy: Techniques, Instrumentation and Applications," D.L. Andrews, ed., VCH Publishers, Inc., New York.

Johnson, J.R., Slotterback, T.J., Pratt, D.W., Janda, K.C., and Western, C.M., 1990, *J. Phys. Chem.* 94:5661.

Kanamori, H., and Hirota, E., 1987, *J. Chem. Phys.* 86:3901.

McIlroy, A., and Nesbitt, D.J., 1989, *J. Chem. Phys.* 91:104.

McIlroy, A., and Nesbitt, D.J., 1990, *J. Chem. Phys.* 92:2229.

Miller, T.A., and Bondybey, V.E., 1983, The Jahn-Teller effect in benzenoid cations, *in*: "Molecular Ions: Spectroscopy, Structure and Chemistry," T.A. Miller and V.E. Bondybey, eds., North-Holland.

Mito, A., Sakai, J., and Katayama, M., 1984, *J. Mol. Spect.* 103:26.

Morgan, R.A., University of Bristol, unpublished results.

Northrup, F.J., and Sears, T.J., 1992, *Ann. Rev. Phys. Chem.* 43:127.

Pfab, J., 1992, Laser-induced fluorescence spectroscopy, *in*: "Applied Laser Spectroscopy: Techniques, Instrumentation and Applications," D.L. Andrews, ed., VCH Publishers, Inc., New York.

Rensberger, K.J., Jeffries, J.B., Copeland, R.A., Kohse-Höinghaus, K., Wise, M.L., and Crosley, D.R., 1989, *Appl. Optics.* 28:3556.

Riedle, E., Neusser, H.-J., and Schlag, E.W., 1983, *Farad. Disc. Chem. Soc.* 75:387.

Romanini, D., and Lehmann, K.K., 1993, *J. Chem. Phys.* 99:6287.

Struve, W.S., 1989, "Fundamentals of Molecular Spectroscopy," John Wiley and Sons, New York.

Vasudev, R., Zare, R.N., and Dixon, R.N., 1984, *J. Chem. Phys.* 80:4863.

Verma, K.K., Bahns, J.T., Rajaei-Rizi, A.R., Stwalley, W.C., and Zemke, W.T., 1983, *J. Chem. Phys.* 78:3599.

Wheeler, M.D., University of Bristol, unpublished results.

INFRARED LASER SPECTROSCOPY OF SHORT-LIVED
ATOMS AND MOLECULES

Paul B. Davies

University of Cambridge
Department of Chemistry
Lensfield Road
Cambridge, CB2 1EW
England

INTRODUCTION

The infrared spectroscopy of free radicals and ions at Doppler limited resolution has produced impressive results over the past two decades. Initial successes in measuring the absorption spectra of transient species, primarily neutral free radicals, was due to highly sensitive techniques like Laser Magnetic Resonance (LMR), (Evenson, 1981). In infrared LMR a ro-vibrational or fine structure transition is tuned through coincidence with a fixed frequency laser line using a magnetic field. The laser sources are gas lasers using different isotopic forms of CO and CO_2. The sensitivity of the technique primarily derives from an intracavity absorption configuration. Amongst the notable discoveries with LMR has been, for example, the first gas phase spectroscopy of the OF free radical, (McKellar, 1979). LMR has undergone continual development in the meantime with particular emphasis on extending the wavelength coverage of the laser. The CO laser is now line tunable from 1200 to 2100 cm^{-1} and from 2500 to 3800 cm^{-1}, (Bachem et al., 1993). Nevertheless two drawbacks remain with LMR. Firstly, the transition must lie close to a laser line even at low J when the Zeeman effect is largest. Secondly, the magnetic field effects can lead to complicated Zeeman patterns - even for diatomic free radicals.

Different types of tunable cw infrared laser have given new impetus to the field. Doppler limited resolution can still be routinely achieved with these sources. The principal lasers now used for absorption spectroscopy in a wide variety of chemical sources such as discharges and beams, are the colour centre laser (CCL), the difference frequency laser, and the lead salt diode laser (TDL). This article describes some representative results using these lasers and in particular the tunable diode laser. The basic principles and characteristics of the lasers are described in outline in the next section.

An Introduction to Laser Spectroscopy
Edited by David L. Andrews and Andrey A. Demidov, Plenum Press, New York, 1995

SPECTROSCOPIC SOURCES

The three cw tunable laser sources currently in use for high resolution spectroscopy are all based on solid state effects. A colour centre (or F-centre) occurs in an alkali halide crystal when a lattice anion is replaced by an electron. Three types of colour centre derived from this basic structure have been used for colour centre lasers namely, F_A, F_B and F_2^+ centres. A full description of these devices is given in the excellent article by Mollenauer and Olson (1975). Colour centre lasers are pumped by Kr^+ or Ar^+ lasers. The output of the laser is varied by placing the cooled crystal inside a cavity containing a grating which can wavelength tune the output. Colour centre lasers are available from 0.8 to 3.6 μm (2700-12500 cm^{-1}). Output powers up to 15 mW have been reported and single mode scans of 10 cm^{-1} have been achieved in a colour centre laser spectrometer at a resolution of 10^{-4} cm^{-1}.

In difference frequency generation two visible lasers, one of which is tunable, are mixed in a nonlinear crystal such as $LiNbO_3$. The usual combination of lasers is a dye laser with an Ar^+ laser. Difference frequency lasers have the drawback of low power, usually below 100 μW. They are tunable between 2500 and 5000 cm^{-1} and single scans of 10 cm^{-1} have been obtained. To achieve the optimum resolution from these devices the visible lasers need to be frequency stabilised. The reader is referred to the original description of the difference frequency laser built by Pine (1974) for more details.

The lead salt or diode laser covers the widest spectral region in the infrared and can operate cw as low as 500 cm^{-1}. However, many devices of different composition are required to cover even small regions of the spectrum. Despite over two decades of development diode lasers are still far from perfect as spectroscopic sources. They often have poor mode structure such as unpredictable multimode behaviour. Modes are usually separated by several wavenumbers leaving as much as 75% of the spectrum uncovered, i.e. several diodes are required for complete coverage of the same spectral region. The optimum performance from diode lasers is above 2000 cm^{-1}, where the detectors are also more sensitive than at longer wavelengths. Attempts to improve diode laser performance are continuing. For example, optical feedback has been used to substantially narrow their linewidth and stabilise their frequency, (Muntz et al., 1992). Another important change has been to raise the cryogenic temperatures over which they operate, i.e. to liquid nitrogen temperatures (77 K) and above. This has removed a major source of noise associated with mechanical vibrations of the cooling system used for liquid helium cooled lasers. An important growth area in the field of diode lasers has been the rapid development of communication diodes as spectroscopic sources in the near infrared e.g. GaAlAs at 0.78 μm, (Bowring et al., 1994). These lasers are technologically more reliable and operate near room temperature. Single mode scans of 10 cm^{-1} are the rule rather than the exception. In addition they are powerful enough to use in combination with fm spectroscopy which greatly increases (by up to a factor of 10) the sensitivity of IR absorption spectrometers.

GENERATING TRANSIENT MOLECULES

Discharge-Flow Systems

This is one of the earliest methods used to form free radicals, particularly the longer lived species. The transient molecule is generated in a flowing gas absorption cell fitted with multiple pass mirrors (White cell) to increase the effective pathlength. The transient can either be formed in a microwave discharge in a side arm of the cell and rapidly pumped into the IR absorption region or formed in an atom-molecule reaction within the cell itself. With the latter method reactive atoms like H, O and F are used, for example, to form the NF free radical in the reaction $H + NF_2 = HF + NF$, (Davies et al., 1981). Figure 1 shows the experimental arrangement used to measure the ν_2 fundamental band of CF_2 which is a long lived transient molecule, (Qian and Davies, 1995). It was formed by a 2450 MHz microwave discharge in a side-arm of the cell using a variety of fluorocarbon precursors. Recently there has been interest in replacing the more indiscriminately reactive atoms like O and F with metal atoms. A major drawback of the former is very little chemical control over the reaction products. Metal atoms like Na can be used to abstract a Br atom from halocarbons to produce hydrocarbon radicals. (Reaction of these radicals with Na is thermodynamically unfavourable). This promises to be a useful route to forming larger radicals which would not survive in the presence of more reactive oxygen and fluorine atoms.

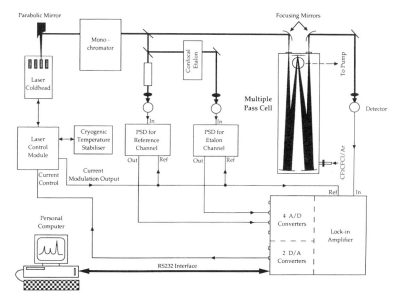

Figure 1. Schematic layout of diode laser spectrometer for detecting transient molecules. The multiple pass cell is shown in the top right hand section of the figure.

Photolytic Methods

Laser photolysis is now widely used to generate free radicals for IR spectroscopy. The commonest experimental arrangement uses an excimer laser to photolyse a low pressure gaseous precursor within a multiple pass cell, very similar to that described above (Figure 1). The photolysis laser usually passes axially along the cell; a common problem is separating the powerful photolysis laser beam from the infrared beam path and sensitive infrared detectors. The pulsed nature of the excimer laser has been exploited in this type of experiment in two ways. Firstly, gated electronic detection allows the infrared absorption to be sampled before and during the laser pulse at a high repetition rate as the cw infrared laser is scanned slowly through the spectrum. A recent example of this type of IR spectroscopy is the work of Sears and co-workers (1992) on the free radical HOCO (DOCO) formed by 193 nm photolysis of acetic acid. A second advantage of gated detection is that it can be used to probe dynamic behaviour of transient molecules formed in the photolysis by varying the delay time between the photolysis pulse and acquisition of the IR absorption spectrum.

Kinetic behaviour of the SiH_2 radical formed by photolysis of phenylsilane has been investigated this way, (Yamada et al., 1989).

A less frequent use of the excimer laser in conjunction with IR spectroscopy is as an ablation source. In this case the laser is used to volatilise normally involatile substances like graphite. Heath and Saykally (1991) have produced several low atomic weight carbon cluster molecules using this method and studied their structure by diode laser absorption spectroscopy. Although a multiple pass cell was used, the transient species were formed in a molecular beam rather than as a low pressure, ambient temperature gas. In this type of application, and in more conventional cells, the indiscriminate nature of powerful laser photolysis may produce a variety of transient species which could potentially present a problem in spectral analysis.

Lastly, non-laser photolysis has been used to form radicals particularly in kinetic spectroscopy applications. The absorption cell is a conventional White-type arrangement enclosed by UV photolysis black lamps, (Cattell et al., 1986). The HO_2 radical has been investigated in this way using either a vibration-rotation line of one of its fundamental bands or an infrared electronic transition $(^2A''\leftarrow{}^2A')$ to monitor its reactions. The radical itself was produced by photolysis of a mixture of Cl_2, H_2 and air.

Discharges

Electric discharges of various types are singularly important as sources of molecular ions for infrared spectroscopy. Following the pioneering work of Oka (1980) on the infrared laser absorption spectroscopy of H_3^+ in a dc glow discharge many other cations and anions have been studied in discharges. In the conventional dc glow discharge the ions are not uniformly distributed either along the axis or radially. Two modifications of the simple two electrode glow discharge have been made to enhance ion production by extending the negative glow region. In a hollow cathode cell the negative glow is extended by enlarging the cathode to fill the whole cell. Only a small pin anode is used. The most common design derives from that of Van den Heuvel and Dymanus (1982). The second method, less frequently used in the IR, is to enclose the discharge cell in a solenoidal magnetic field. The axial magnetic field confines charged species along the cell axis and reduces wall removal.

Although electric discharges are fruitful sources of ions and free radicals they are nevertheless indiscriminate sources. Discharges in polyatomic gases can give rise to a range of atoms, radicals and ions in ground and excited states. The IR spectra of such discharges can be extremely complicated and a method of species selection before spectroscopic detection is desirable. Cao et al. (1993) have developed an IR spectrometer which uses mass selection before spectroscopy. The ion of interest, SiH_7^+, was mass selected from a high pressure discharge. The interaction of SiH_7^+ with the IR laser occurred in a molecular beam chamber. The conventional glow discharge is not easily adapted for this type of experiment. Instead they used a Corona discharge in a 1000 Torr mixture of SiH_4 and H_2 (SiH_4: H_2, 1:30,000). The discharge was struck from a pin electrode in the high pressure gas mixture, through a 70 μm nozzle which formed a molecular beam of ionic species. The IR spectrum of SiH_7^+ was detected using a second mass selection system after the interaction region. This was tuned to the mass of the ion SiH_5^+ which is formed when SiH_7^+ is vibrationally predissociated on absorbing a photon: $SiH_7^+ + v_{IR} = SiH_5^+ + H_2$. This type of experiment is likely to become increasingly important for infrared spectroscopy of cluster ions.

One of the most significant advances in IR spectroscopy of discharges occurred when a method of distinguishing ions from neutral species was developed. This is the technique called Velocity Modulation, (Gudeman et al., 1983) which is used with ac discharges. In an ac discharge the infrared absorption of ions is red and blue shifted with respect to the laser frequency due to the change in direction of the ions as the polarity of the electrodes switches. Detection with a lock-in amplifier tuned to the ac discharge frequency (f) gives a first derivative absorption signal. In the most favourable cases velocity modulation detection completely discriminates against neutral transient molecules formed in the discharge. Since the ions constitute a very small proportion of the total concentrations present this is a very useful ion specific technique. Velocity modulation spectrometers achieve sensitivities of $\sim 10^9$ ions cm^{-3}. In addition the phase of the first derivative signal contains useful information; anions and cations produce signals of opposite phase. The ac discharge has one further advantage. The discharge switches on twice during the discharge cycle, i.e. at twice the discharge frequency (2f). If the detection system is now tuned to 2f all transient molecules are detected, provided they have IR allowed transitions, with a zeroth order (direct absorption) line profile. The switching on/off of the discharge is analogous to mechanical chopping of the laser beam at 2f (which also produces a zeroth order line shape). Selection of the ac frequency is important in these experiments. A long lived transient would not be efficiently detected by this population or concentration modulation method if the ac frequency were too high, and generally the inverse of the discharge frequency should be of the same order of magnitude as the lifetime of the transient species. Nevertheless, in principle it should be possible to probe the IR spectra of an ac discharge for ions or neutrals by simply switching from 1f to 2f detection.

EXAMPLES

Atoms

Many examples of infrared laser spectra of atoms have been reported using electric discharges to produce excited atomic states. The transitions often have large oscillator

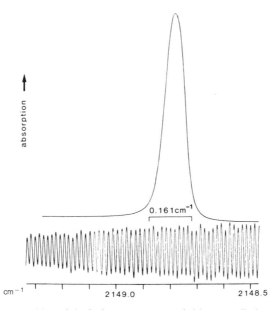

Figure 2. The n = 7 ← 5 transition of the hydrogen atom recorded in an ac discharge through H$_2$ and He. The interference pattern from a germanium etalon, used for relative calibration, is shown below the absorption line.

strengths and can be measured at high resolution with good signal to noise ratios revealing fine structure and hyperfine structure splittings. The simplest example is the hydrogen atom itself. Highly excited states are easy to produce in discharges in H_2 and an example is shown in figure 2, (Davies et al., 1988). This single absorption feature was recorded in a dc discharge without discharge modulation and corresponds to 15% absorption of the laser intensity. Figure 3 shows three transitions in the infrared Rydberg-Rydberg spectra of atomic oxygen, (Brown et al., 1984). Rydberg states of this atom are easily populated in an ac discharge in oxygen and infrared laser spectra have been recorded of levels within 5000 cm^{-1} of ionisation (I.P. ~ 110,000 cm^{-1}). The splitting shown in figure 3 arises from fine structure. The same transition was detected in emission in a laser dissociated oxygen plasma in which the linewidths were 12-50 cm^{-1} and fine structure was not resolved. The assignment of these Rydberg-Rydberg transitions is relatively straightforward. Term values for the energy levels involved are either available from spectroscopy in the visible or UV region or can be calculated from polarisation theory. The same experimental and theoretical approach has been used for many other atoms. In fact atomic lines appear in many molecular discharges and can complicate spectral analysis. On the other hand although atomic transitions are easy to observe using population modulation relatively little effort has been expended in understanding the dynamics of discharges using IR atomic transitions as probes.

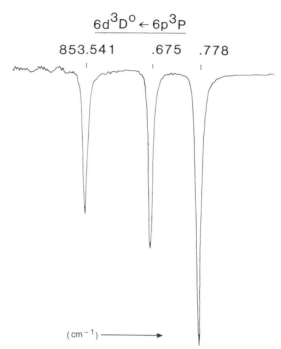

$$6d\,^3D^o \leftarrow 6p\,^3P$$

853.541   .675   .778

(cm^{-1}) ⟶

Figure 3. Three fine structure components of an OI transition recorded in absorption in a 50 kHz discharge in a mixture of O_2 (20 mTorr) and He (160 mTorr). The two higher cm^{-1} lines contain further unresolved fine structure from the ^3D state.

Free Radicals

Over the past few years several free radicals have been positively identified in the gas phase for the first time using IR laser spectroscopy. One of the most important is HOCO which has been produced photolytically and by atom-molecule reaction. Both *ab initio*

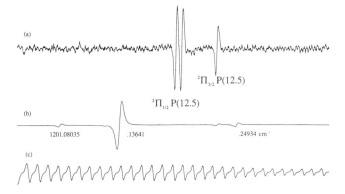

Figure 4. Diode laser absorption spectrum of two rotational transitions of the PO radical. Spectra (b) and (c) are calibration lines and the etalon fringe pattern respectively.

calculations and matrix IR spectroscopy have made important contributions towards analysing the gas phase IR spectra of HOCO. The first gas phase IR spectrum was measured by Sears et al. (1992) who detected many lines in the v_2 (C=O stretch) fundamental of the *trans* isomer. The radical is planar and the gas phase v_2 band origin is blue shifted by 9 cm^{-1} from the matrix value. More recently the v_1 (O-D stretch) fundamental in t-DOCO around 2684 cm^{-1} was detected using a difference frequency laser, by Petty and Moore (1993). In the latter study the radical was formed by 193 nm photolysis of deuterated acrylic acid with an excimer laser. This band is also blue shifted (by ~1%) from the condensed phase band origin. Comparison of the experimental B values with *ab initio* calculations points unambiguously to the *trans* isomer. The planarity of the radical was supported by comparing the inertial defects (0.071 amu Å^2, $v = 0$; 0.070 amu Å^2, $v_1 = 1$) with that of formaldehyde (0.058 amu Å^2).

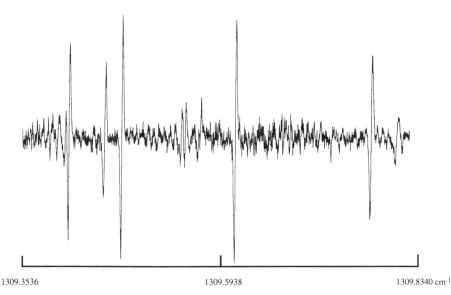

1309.3536   1309.5938   1309.8340 cm^{-1}

Figure 5. Diode laser absorption spectrum of the PO$_2$ radical, formed in the reaction of oxygen atoms with the vapour of white phosphorus.

Recently we have examined the transient species formed in the reaction of atomic oxygen with the vapour of white phosphorus in a discharge flow cell. The experimental arrangement is shown in figure 1. The reaction produces PO and PO_2 radicals as well as other transient species like P_2O and P_4O. Figure 4 shows two rotational transitions in the fundamental band of the PO radical in its ground $^2\Pi$ state. The doubling in the $^2\Pi_{1/2}$ transition is due to Λ doubling, which is not resolved in the $^2\Pi_{3/2}$ component. The spectrum in Fig.5 is part of the v_3 antisymmetric stretching fundamental of PO_2, also recorded using a diode laser in the $O + P_4$ reaction. Both PO and PO_2 were detected using source modulation of the diode laser which does not distinguish between paramagnetic and non-paramagnetic species. However, for these examples sufficient information was available from other studies, Butler et al. (1983); Kawaguchi et al. (1985); Hamilton (1987) to be certain of the identity of the carriers of the spectra.

The silyl radical, SiH_3, has been widely studied using diode laser spectroscopy. Part of the motivation for interest in it is that IR detection is a convenient way of monitoring it in the silane plasmas used to form amorphous silicon for semiconductor applications. It was first detected by Yamada and Hirota (1986) using a dc glow discharge in silane, and diode lasers with Zeeman modulation of the discharge cell. Their study of the v_2 fundamental showed that the radical was non-planar with two sub-bands with origins near 721 and 727 cm^{-1}. The v_3 degenerate stretch of SiH_3 has recently been detected in a discharge in phenyl silane, using diode laser absorption spectroscopy and a multipass cell, Sumiyoshi et al. (1994).

Rydberg States

Excited electronic states of molecules as well as atoms are often produced in both dc and ac discharges. Molecular Rydberg states formed in this way have been studied by IR laser and Fourier transform spectroscopies. Rydberg-Rydberg spectra of H_2 have been extensively investigated throughout the infrared region in absorption and emission. These IR spectra often appear in systems containing hydrogen or a hydrogen containing precursor in the discharge. The spectroscopy of Rydberg-Rydberg transitions in H_2 was initiated by Herzberg and Jungen (1982) who used FTIR to examine, for example, the 4f-5g and 5g-6h systems near 2500 and 1350 cm^{-1} respectively. A recent example of the IR laser spectroscopy of high l transitions is the diode laser study of the 7h-6g and 7i-6h systems which partly overlap each other near 810 cm^{-1}. The spectra have no obvious patterns to aid assignment. Instead assignment is achieved by calculating the ro-vibrational levels as precisely as possible using *ab initio* techniques. Figure 6 shows part of the spectrum where the dominant lines are due to 7h-6g transitions, (Basterechea et al., 1994). The stick diagram below the spectrum shows the calculated pattern and relative intensities. This is an unorthodox method of assignment for a spectroscopist but the theoretical models work well and predictions are reliable for hydrogen within certain limitations. The calculations used in this case were based on the multipole-polarisation model of the Rydberg molecule. This model is accurate for high l and n states but gets progressively less precise at low l (n » l) when the Rydberg electron penetrates the region of the H_2^+ core. The main input parameters into the calculation, in addition to the Rydberg energy (R/n^2) and ionisation energy (I), are the isotropic and anisotropic polarisability of the core, the quadrupole and higher moments and the ro-vibrational energy of the H_2^+ core. It is now possible to calculate the spectra with a precision matching the experimental accuracy of 0.003 cm^{-1}. There is considerable interest in extending this type of spectroscopy to Rydberg states of other diatomics including those with dipolar cores, for example, NO.

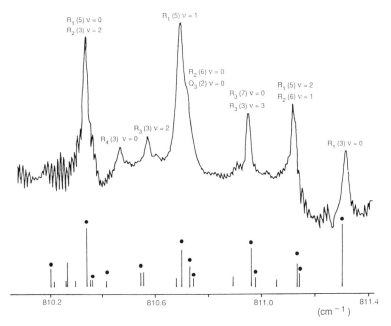

Figure 6. Diode laser absorption spectrum of Rydberg-Rydberg transitions in H_2. The more intense lines are due to the 7h-6g system shown as (•) on the stick diagram below the spectrum. Other lines are due to 7i-6h transitions.

Ions

Two examples of ions studied by IR laser spectroscopy will now be described. They illustrate the use of both velocity modulation and hollow cathode techniques. The latter technique has proved to be useful for forming protonated organic molecules. In his study of CH_3CNH^+ Amano (1992) used a (cooled) hollow cathode discharge in methyl cyanide and hydrogen. The cathode voltage was modulated at 10 kHz (population modulation). To start the search for spectra *ab initio* calculations of the band origin of the v_1 (N-H stretch), 3550 cm^{-1}, were used. The experimental spectrum had the expected features of a parallel band of a symmetric top with assigned P and R branches. The cation retains the C_{3v} symmetry of its neutral symmetric top precursor and is protonated at the N atom. The band origin was located at 3527.2887 cm^{-1} which is satisfactorily close to the calculated value and illustrates the usefulness of precision calculations for high resolution spectroscopy. The ground state rotational constant was also precisely determined experimentally and may be useful in future searches for the rotational spectrum of the ion in the laboratory or in interstellar space. Amano pointed out that the calculated value for its dipole moment, 1.25 D, is large enough to expect laboratory observation of rotational spectra whilst the abundance of CH_3CN in interstellar space suggests that CH_3CNH^+ may be present in the interstellar medium.

The SiH_3^+ cation, like SiH_3 neutral, has attracted spectroscopic attention because of its potential importance in silane plasmas. Prior to its identification by high resolution IR spectroscopy the ion was thought to have a planar structure with two of its four fundamental frequencies lying below 1000 cm^{-1}. Both of these modes, the v_2 out-of-plane bend and the v_4 degenerate in-plane bend, have been detected and studied by diode laser-velocity modulation spectroscopy, (Davies and Smith, 1994). The starting points in searching for SiH_3^+ spectra were the *ab initio* calculations of the band origins and the known ion-molecule chemistry in silane plasmas. The first spectrum discovered was the J = K, Q-branch of the v_2 fundamental band shown in figure 7. This simple spectrum provided two vital pieces of information about the ion giving rise to it. Firstly, it had a three-fold symmetry axis since the spectrum shows the expected 2:1:1:2 intensity alternation for K = 3n and K $\neq$ 3n rotational levels, i.e. there are three equivalent I = ½ nuclei. Secondly, the spacing of the rotational components in figure 7 yields directly a value of ΔB ($B_{v_2=1}$ − $B_{v=0}$). This was in excellent agreement with the result of an *ab initio* calculation which showed a strong Coriolis interaction between the upper vibrational levels of the v_2 and v_4 bands. Other sections of the v_2 band were then assigned in a step by step procedure beginning with low J members of the P and R branches.

The assignment of the v_4 perpendicular type band required a different approach. The most intense lines in its spectrum were predicted to be the ^rR and ^pP lines which are separated by ~ 6 cm^{-1}. Once enough of these had been identified a global fit of both bands was undertaken to produce ground and excited state constants. An iterative procedure of prediction and fitting then allowed more lines to be assigned and included in the least squares fit. Figure 8 is a stick diagram drawn up after both bands had been analysed. It shows that some band structure could be expected for the v_2 fundamental but virtually none for the v_4 band. This figure should convey some of the difficulties encountered in unravelling partially overlapped bands at high resolution whilst hampered by the limited signal to noise ratios achievable when studying a very short lived ion. The vibrational band

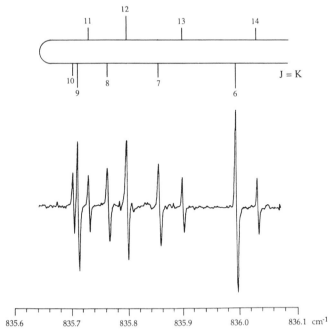

Figure 7. Diode laser absorption spectrum of SiH_3^+ recorded using velocity modulation spectroscopy. The intensity variation of the lines in this J = K, Q-branch of the v_2 fundamental shows the effect of nuclear spin statistical weights.

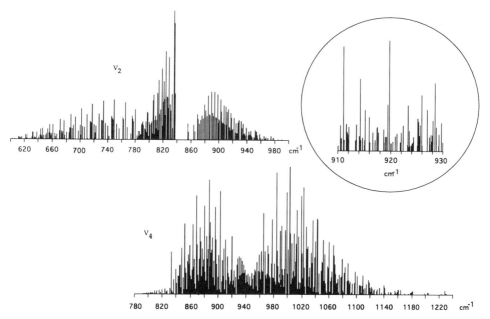

Figure 8. Simulated spectrum of the v_2 and v_4 fundamental bands of SiH_3^+ at a rotational temperature of 570 K. Inset is a small region of the spectrum where both bands overlap resulting in a high density of lines.

origins are probably the most interesting parameters to be derived from the analysis because they can be compared directly with theoretical values. The spectroscopically determined quantities are $\tilde{v}_0(v2) = 838.0669$ cm^{-1} and $\tilde{v}_0(v4) = 938.3969$ cm^{-1}. These are in remarkably good agreement with the *ab initio* result of Botschwina and Oswald (1992), who obtained 842 and 941 cm^{-1} respectively.

CONCLUDING REMARKS

The increasing size and complexity of the transients now being studied is leading to increasing sophistication in methods of production of molecules, particularly ion complexes. The drive for ever higher sensitivity has led to the development of high frequency modulation techniques particularly in the near infrared. Limiting detectable absorption has been extended from 10^{-5} to 10^{-7} using fm methods. Applications of tunable IR laser spectroscopy are mainly to be found in atmospheric chemistry (both for chemical kinetic studies and trace gas monitoring) and in chemical vapour deposition research. Future IR laser spectroscopy promises to become even more closely allied to two associated areas of research: firstly, matrix isolation spectroscopy which can indicate plausible species for gas phase investigation as well as approximate values for their fundamental band frequencies; secondly, *ab initio* calculations which are giving accurate predictions of vibrational frequencies and intensities for ions and free radicals.

ACKNOWLEDGMENTS

I am grateful to Hai-bo Qian for kindly supplying figures 1, 4 and 5.

REFERENCES

Amano, T., 1992, *J. Mol. Spectr.* 153:654.

Bachem, E., Dax, A., Fink, T., Weidenfeller, A., Schneider, M., and Urban, W., 1993, *Appl. Phys.* B57:185. Basterechea, F. J., Davies, P. B., Smith, D. M., and Stickland, R. J., 1994, *Mol. Phys.* 81:1435.

Botschwina, P., and Oswald, M., 1992, *J. Chem. Phys.* 96:4044.

Bowring, N. J., Li, D., and Baker, J. G., 1994, *Meas. Sci. Technol.* 5:1313.

Brown, P. R., Davies, P. B., and Johnson, S. A., 1987, *Chem. Phys. Lett.* 133:239.

Butler, J. E., Kawaguchi, K., and Hirota, E., 1983, *J. Mol. Spectr.* 101:161.

Cao, Y., Choi, J-H., Haas, B-M., Johnson, M. S., and Okumura, M., 1993, *J. Phys. Chem.* 97:5215.

Cattell, F. C., Cavanagh, J., Cox, R. A., and Jenkin, M. E., 1986, *J. Chem. Soc. Faraday II* 82:1999.

Davies, P. B., Hamilton, P. A., and Okumura, M., 1981, *J. Chem. Phys.* 75:4294.

Davies, P. B., Guest, M. A., and Johnson, S. A., 1988, *J. Chem. Phys.* 88:2884.

Davies, P. B., and Smith, D. M., 1994, *J. Chem. Phys.* 100:6166.

Evenson, K. M., 1981, *Disc. Faraday Soc.* 71:7.

Gudeman, S. C., Begemann, M. H., Pfaff, J., and Saykally, R. J., 1983, *Phys. Rev. Lett.* 50:727.

Hamilton, P. A., 1987, *J. Chem. Phys.* 86:33.

Heath, J. R., and Saykally, R. J., 1991, *J. Chem. Phys.* 94:1724.

Herzberg, G., and Jungen, Ch., 1982, *J. Chem. Phys.* 77:5876.

Kawaguchi, K., Saito, S., Hirota, E., and Ohashi, N., 1985, *J. Chem. Phys.* 82:4893.

McKellar, A. R. W., 1979, *Can. J. Phys.* 57:2106.

Mollenauer, L. F., and Olson, D. H., 1975, *J. Appl. Phys.* 46:3109.

Müntz, M., Schaefer, M., Schneider, M., Wells, J. S., Urban, W., Schiessel, U., and Tacke, M., 1992, *Opt. Comms.*, 94:551.

Oka, T., 1980, *Phys. Rev. Lett.* 45:531.

Petty, J. T., and Moore, C. B., 1993, *J. Chem. Phys.* 99:47.

Pine, A., 1974, *J. Opt. Soc. Am.* 64:1683.

Qian, H.-B., and Davies, P. B., 1995, *J. Mol. Spectr.* 169:201.

Sears, T. J., Fawzy, W. M., and Johnson, P. M., 1992, *J. Chem. Phys.* 97:3996.

Sumiyoshi, Y., Tanaka, K., and Tanaka, T., 1994, *Appl. Surf. Sci.* 79/80:471.

Van den Heuvel, F. C., and Dymanus, A., 1982, *Chem. Phys. Lett.* 92:219.

Yamada, C., and Hirota, E., 1986, *Phys. Rev. Lett.* 56:923.

Yamada, C., Kanamori, H., Hirota, E., Nishiwaki, N., Itabashi, N., Kato, K., and Goto, T., 1989, *J. Chem. Phys.* 91:4582.

HIGH SENSITIVITY PICOSECOND LASER SPECTROSCOPY

W. Jeremy Jones

Department of Chemistry
University of Wales Swansea
Singleton Park
Swansea, SA2 8PP

INTRODUCTION

Studies of fast spectroscopic events in atomic and molecular systems occurring on the picosecond or sub-picosecond time scales generally require short pulse excitation sources to create a population perturbation among discrete energy states and a means of monitoring the perturbation process or the recovery of the system towards its equilibrium conditions. In recent years excitation methods have benefited dramatically from the development of picosecond and femtosecond lasers operating in the visible regions of the spectrum, and derivatives of these excitation sources operating in the ultraviolet region (by frequency doubling and trebling) and in the infrared region (by difference frequency generation). Following hand-in-hand with these developments in short pulse sources of electromagnetic radiation have been improved fast detection methods, such as streak cameras, fast response photomultipliers and photodiodes, and pump-probe spectroscopic methods, which permit these perturbations to be followed to the picosecond and in some cases to the femtosecond time domains. The extent to which such experimental methods may be employed to monitor transient spectroscopic events, however, depends not only on the time response of the overall system but also on the sensitivity of the detection methods to the perturbation created. In seeking to investigate the applicability of these methods to chemical systems there is, not surprisingly, a limitation set by the Heisenberg uncertainty principle between the temporal characteristics of the radiation source employed and the quantised energy states which may be interrogated. This limitation is most readily seen by an appreciation of the transform relationship which exists between the time duration of a pulse of laser radiation and the spectral width of the band of radiation of which it is comprised. The time-frequency relationship $\Delta v \Delta t \geq \frac{1}{2}\pi$ implied by the uncertainty principle, Δt and Δv being the full widths at half maximum height of the temporal pulse and the measurable frequency spectrum of the radiation beam, implies that shorter pulses can only be generated at the expense of the spectral purity of the radiation source. The precise form of this relationship depends among other factors on the spectral profile of the radiation source creating the laser pulse. For transform limited gaussian laser pulses of various durations

An Introduction to Laser Spectroscopy
Edited by David L. Andrews and Andrey A. Demidov, Plenum Press, New York, 1995

the following Table 1 displays the associated spectral bandwidth, expressed in convenient wavenumber (cm^{-1}) units (Ippen and Shank, 1977; Bradley, 1977)

Table 1. Transform-related Spectral and Temporal Linewidths of Laser Sources

Pulse width, Δt	150 ps	15 ps	1.5 ps	15 fs
Spectral width, $\Delta v/cm^{-1}$	0.1	1	10	1000

The data presented in this table show clearly that the shortest femtosecond events which can be investigated by current technology can only be examined by radiation sources which cover a broad range of excitation wavelengths. Such radiation sources, however, are of little value for the study of the spectra of atoms, or molecular species possessing closely spaced transitions, such as are found in vibrational structures of molecular spectra, whether in the infra red region or accompanying electronic transitions in electronic spectra. Selective excitation of individual features with discrete structure in the range 1 - 10 cm^{-1} can only be accomplished by the use of picosecond sources, thereby limiting the time response of the fastest processes of such features which may be studied to the picosecond regime. Even in the study of broad electronic spectra, such as are found for many molecular species in the liquid state, quantitative information on absorption coefficients, and hence on the rate of photon extraction from the radiation source, generally requires the use of lasers whose spectral line width is narrow in comparison with the spectral transitions under investigation. Because of these restrictions on the applicability of femtosecond lasers, spectroscopic studies of fast processes, other than the temporal studies of very short duration processes associated with broad spectral transitions, are more appropriately studied by picosecond rather than femtosecond laser systems.

PUMP-PROBE SPECTROSCOPY

With the development of picosecond and femtosecond laser systems in recent decades extensive studies have been carried out on a wide variety of ultrafast processes in Physics, Chemistry and Biology. Typical of such studies are investigations of various non-linear optical processes, vibrational dephasing and vibrational relaxation processes, picosecond and femtosecond relaxation in visual pigments in biological systems, intermolecular and intramolecular energy transfer processes in solids, liquids and gases, *etc*. The range of applications is extensive and for a detailed coverage of the work that has been promoted and the techniques that have been developed the reader is referred to extensive reviews on the subject of ultrafast spectroscopy, such as that of Shapiro (ed; 1977), Demtröder (1982) or Fleming (1986).

In so far as the study of the state perturbation created by the excitation, or pump, laser is concerned, because of the limited temporal resolution of high sensitivity photomultiplier detectors, and the low sensitivity and lack of a linear response of fast streak cameras, most quantitative investigations of picosecond/femtosecond optical processes have employed variations of the pump-probe spectroscopic method. In this technique two individual sets of laser pulses of comparable time duration are incident upon the same region of space of an absorbing sample. Absorption of photons from the initially incident pump pulse creates a population perturbation between the two states coupled by the radiation, a certain proportion of the atoms/molecules initially present in the ground state being excited to the higher energy state during the duration of the pulse. Recovery to the equilibrium condition following this perturbation occurs on a time scale which is determined by the internal energy decay routes open to the excited state species. Since the transmission of the probe laser through the sample is influenced by the populations of the states perturbed by the pump laser, the recovery of the pump-perturbed populations towards equilibrium can be monitored by delaying the probe pulse in time following pump excitation. Such a time delay is readily introduced by increasing the optical path traversed by the probe, *e.g.* by moving a retro-reflecting corner-cube prism (since the optical path delay introduced by moving a corner-cube retro-reflector is twice the displacement, a 1 mm movement introduces a time delay of 6.7 ps).

It does not follow that pump and probe lasers need to be of the same wavelength, and indeed there are many circumstances where it is highly advantageous that they be different. Such circumstances include the need to separate pump and probe-lasers post-sample so that the change of the transmission of the probe laser can be monitored in the absence of stray light from the pump beam. With different wavelength pump and probe lasers this separation of the beams after the sample is readily accomplished by a dispersive optical element, such as a diffraction grating, or a band pass filter. For pump and probe lasers of the same wavelength such a procedure is not possible and separation is generally accomplished for coaxial forward propagating beams of orthogonal polarisations by means of optical polarisers. However, while this procedure is quite satisfactory for the gas phase, it is not so suitable for condensed phase species where orientational decay processes of large chromophores take place on a relatively long time scale. Different wavelength pump and probe lasers avoid many of these difficulties but can only be used where the probe lasers can interrogate one or other of the two states perturbed by the pump. If the probe laser lies in the absorption band of the ground state species it interrogates the hole in the ground state concentration created by the absorptions of pump photons. For some samples, however, the probe laser may additionally experience absorption from the excited state species created, or stimulated emission as the excited state species is induced to emit its excitation energy and revert to the ground state. In certain samples many of these processes may occur at the same time and often under the control of the experimenter so that pump-probe spectroscopy represents a family of techniques providing information which is spectral, temporal and spatial in nature.

Since the methods of pump-probe spectroscopy monitor the change of the transmitted probe laser power under the perturbing influence of the pump, it is particularly important to employ radiation detectors capable of monitoring small fractional probe laser power changes. In this respect silicon photodiode detectors are particularly attractive since: (i) they are extremely fast with response times which can approach the picosecond time domain; (ii) they operate linearly over a range of some 8-9 orders of magnitude of incident power, or more; (iii) they can operate at incident light power levels of tens of milliwatts per square centimetre, rendering them capable of responding to intensity fluctuations comparable with the shot noise limit of ~ 1 part in 10^8 for ~ 3 mW power on the detector at 600 nm. The manner in which these detectors may be employed to attain this exceptional level of sensitivity will be described below, together with some examples of the use of these methods in various spectroscopic applications.

Concentration Perturbation

The essence of the pump-probe spectroscopic method can be appreciated from figure 1 which displays the perturbation to the state concentrations of levels 1 and 2 generated by the absorption of photons from a pulse of pump laser radiation. This diagram is drawn schematically for a gaseous absorbing entity with discrete energy states. It is important to appreciate, however, that similar sets of energy states occur in a variety of condensed phase species except that the relevant energy states are broadened substantially through the array of vibrational energy states which may accompany the electronic states of molecules and the level broadenings that occur in the condensed phases as a result of the incessant collisions and interactions which occur in these high density phases. A variety of probe laser wavelengths are displayed in figure 1, demonstrating some of the transitions which may be employed to monitor the perturbation to levels 1 and 2 created by the pump laser.

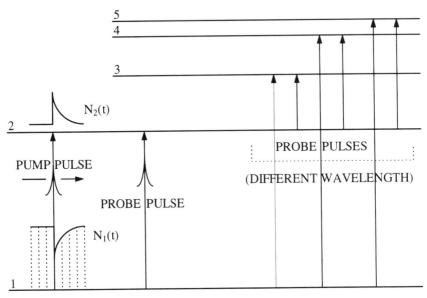

Figure 1. Schematic energy level scheme. It is assumed that the pump laser perturbs the population of level 1 (the ground state) and level 2. Some of the array of probe wavelengths which may be employed to monitor the perturbation created by the pump laser are indicated. The left side of the diagram displays the population perturbation of levels 1 and 2 and their decay with time following pump pulse excitation.

To evaluate the concentration perturbation created, we consider that the picosecond pump pulse which perturbs the state concentrations contains i_p photons in a pulse of picosecond duration. Since it proves to be advantageous to focus pump and probe beams together into the sample it is necessary to evaluate this perturbation in localised zones near the combined focus of the two beams. Suitable quality laser beams necessarily possess axial symmetry and for a gaussian intensity distribution the infinitesimal flux of photons in radial element dr, distance r from the beam axis, is given by (Langley et al., 1986)

$$i_p(r, z) = \frac{4r}{w_p^2(z)} e^{-2r^2/w_p^2(z)} i_p(z) dr .$$ (1)

The length variable along the direction of beam propagation, z, is included to account for (a) the manner in which the beam radius $w_p(z)$ changes in the vicinity of the focus and (b) the attenuation of the pump flux which occurs due to absorption as the pump laser propagates through the sample. The decrease in this photon flux $i_p(r, z)$ due to sample absorption is given by the expression

$$-\frac{di_p(r,z)}{dz} = \sigma_p \, N_1^e \, i_p(r,z), \tag{2}$$

where σ_p is the absorption cross section at the pump wavelength and N_1^e the equilibrium concentration of the species in the ground state. Integration of equation (2) over the cell length L leads to the well known Beer-Lambert relationship which relates the sample absorbance per unit path length to the product of the absorption cross section (extinction coefficient) and the species concentration:

$$\frac{1}{L} \, ln\left(\frac{I_o}{I}\right) = \sigma_p \, N_1^e . \tag{3}$$

In a localised volume element dV (= $2\pi r dr dz$) at position r, radial element dr and thickness dz, the number of absorbed photons expressed by equation (2) monitors also the concentration change in levels 1 and 2 in volume element dV, such that

$$dN_1(r,z) = - \, dN_2(r,z) = \frac{di_p(r,z)}{dV}, \tag{4}$$

yielding for the ground and excited state concentrations immediately following the passage of the pump pulse

$$N_1(r,z) = N_1^e\left\{1 - \frac{2\sigma_p}{\pi w_p^2(z)} . e^{-2r^2/w_p^2(z)} . i_p(z)\right\}, \tag{5a}$$

$$N_2(r,z) = N_1^e \, \frac{2\sigma_p}{\pi w_p^2(z)} \, e^{-2r^2/w_p^2(z)} \, i_p(z). \tag{5b}$$

Equations 5(a) and 5(b) show clearly that the greatest fractional population changes in the ground and excited states occur in the vicinity of the beam focus when the beam area $\pi w_p^2(z)$ is smallest.

Concentration Probing

The concentration perturbation created by the pump pulse is monitored by comparing the transmittance of a probe pulse with and without a preceding pump pulse. It is convenient to consider initially the probe laser to be set to a transition different from the pump laser with absorption cross section σ_{pr}. Considering the probe absorption to take place in volume element dV following perturbation by the pump pulse, we have:

$$-\frac{di_{pr}(r,z)}{dz} = \sigma_{pr} . N_1(r,z) . i_{pr}(r,z) . \tag{6}$$

Allowing for the attenuation of the pump beam with distance z into the absorption cell according to the Beer-Lambert relationship $i_p(z) = i_p(0) \exp(-\sigma_p N_1^e z)$, ($i_p(0)$ being the pump flux incident on the sample), introduction of equations (1) and (5a) into (6) and integration radially at each position z yields

$$i^*_{pr}(L) = i_{pr}(0)\exp(-\sigma_{pr}N_1^e L).\exp\left\{\int_{z=0}^{L}\sigma_p\sigma_{pr}\frac{N_1^e}{\pi w^2(z)}.i_p(0).\exp(-\sigma_p N_1^e z)dz\right\}. \quad (7)$$

It is assumed in the derivation of (7) that pump and probe beams have identical spatial profiles (with $w_p = w_{pr} = w$), the beam radius $w(d)$ of a gaussian beam distance d from the beam focus, where the spot radius is w_0, being given by the expression (Yariv, 1985)

$$w^2(d) = w_0^2\left[1 + \left(\frac{\lambda d}{\pi w_0^2}\right)^2\right]. \quad (8)$$

Introducing equation (8) and suitably transforming the variables allows equation (7) to be integrated for conditions of weak pump and probe absorption, and where the focal zone is fully contained within the cell, to yield

$$i^*_{pr}(L) = i_{pr}(L).\left\{1 + \frac{\pi\,\sigma_p\sigma_{pr}N_1^e\,i_p(0)}{\lambda}\right\}, \quad (9)$$

where λ is the mean wavelength of the pump and probe lasers.

Since in the absence of the preceding pump pulse the transmitted flux of the probe laser is $i_{pr}(L)$, equation (9) shows clearly the magnitude of the increase in the transmitted flux of probe photons, $i^*_{pr}(L)$, as a result of the bleaching of the state population by the preceding pump pulse. Equation (9) is used to quantify what is termed the probe laser "gain", in reality an attenuated absorption, though there is indeed a real increase in the transmitted flux of probe photons through the absorbing sample following pump excitation, defined as

$$G = \frac{i^*_{pr}(L) - i_{pr}(L)}{i_{pr}(L)} = \frac{\pi\,\sigma_p\sigma_{pr}N_1^e\,i_p(0)}{\lambda}. \quad (10)$$

Equation (10) defines the condition where pump and probe lasers interrogate different pairs of energy states, with the lower level in common. Where both pump and probe lasers have the same wavelength, and hence interrogate the same pair of energy states, the excited state population is also relevant and the "gain" expression is modified to (Langley et al., 1985, 1986)

$$G = \frac{2\pi\,\sigma^2 N_1^e\,i_p(0)}{\lambda}. \quad (11)$$

Expressions similar to those of equations (10) and (11) can be derived for conditions where the probe laser is located in regions of stimulated emission or excited state absorption from level 2. Under certain conditions several probe processes may occur simultaneously and although the "gain" expressions under such circumstances are not specified here they are readily derived using the methods described above.

The "gain" expressions cited here have been derived for single pulses of pump and probe laser radiation and as such are ideal for use with mode-locked trains of picosecond pulses, it being assumed that following perturbation by a picosecond pump pulse the perturbation has relaxed back to equilibrium before the next pump pulse arrives at the sample. With separations of ~ 12 ns between individual pulses in a mode-locked train this approach is perfectly adequate for studying relaxation processes with lifetimes in the range 5 ps to 5 ns. In practice even longer lifetimes may be investigated using mode-locked laser systems (Griffiths, 1990), and extensions of the methods to study state lifetimes in the range 10^{-7} - 10^{-2}s are possible using cw lasers and variable frequency modulation methods (Mallawaarachchi et al., 1987; Czarnik-Matusewicz et al., 1988). Since the measured "gain" does not depend on the presence of a fluorescing chromophore for sensitive detection of the excitation process, it is clear that the pump-probe technique represents a powerful method for measuring state lifetimes down to the picosecond domain for all species, whether they fluoresce or not. The methods that have been developed of measuring very small fractional changes of the probe under the perturbing influence of the pump laser, and thereby of attaining high sensitivity for investigating direct absorption processes, are described in detail below.

Concentration-Modulated Absorption Spectroscopy (COMAS)

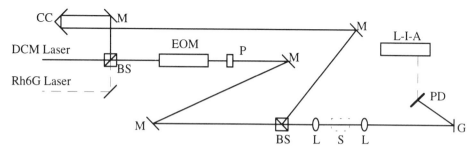

Figure 2. Schematic layout of a typical arrangement for recording pump-probe spectra using mode-locked pump and probe lasers. EOM - Electro-Optic Modulator; P - Polariser; M- Mirrors; BS - Beam-Splitter; CC - Corner Cube; L - Lenses; S - Sample; G - Diffraction Grating; PD - Photodiode Detector; L-I-A - Lock-in-Amplifier.

It is well known that with few exceptions pulsed lasers of the sort that may be employed for temporal studies are intrinsically noisy radiation sources. The pulse-to-pulse reproducibility is rarely better than a few per cent, frequently very much poorer. Two main laser types are in common use for picosecond studies, low repetition rate (~ 10 Hz) lasers of high pulse energy (mJ per pulse) and mode-locked lasers of high repetition rate (~ 80 - 100 MHz) but very low energy (nJ per pulse). The average photon fluxes in a one second time interval (the typical integration time of a spectroscopic system) are comparable for both systems and equate to average powers of several milliwatts. At these average power levels, the maximum average powers that silicon photodiode detectors may assimilate, the photon flux in a 1 second time interval is ~ 10^{16}. Since the intrinsic shot noise associated with such a flux is only ~ 10^8 ($\sqrt{N}$) the pulse-to-pulse reproducibility is some six orders of magnitude greater than this theoretical noise limit. Direct measurements of the transmitted flux of probe photons in a pump-probe experiment, whether individual probe pulses are preceded by a pump pulse or not, are thus likely to exhibit a fluctuation ranging from several per cent to many orders of magnitude lower. Since such fluctuations appear as source noise, against the background of which the signals of interest must be abstracted, the signal-to-noise ratios in this branch of spectroscopy can vary dramatically. It is thus particularly important in optimising signal-to-noise ratios, and thereby enhancing sensitivity, to identify means of approaching the shot-noise-limit.

Accepting a pulse to pulse fluctuation of a few per cent, it is readily appreciated that low repetition rate systems are inherently unsuitable for low noise performance. Rapid repetition rate mode-locked systems, however, are very much more attractive as low noise sources. Apart from the lower individual pulse energies, which avoids saturation of optical transitions and damage to the detector, averaging over many transitions alone during a 1 second interval reduces the noise fluctuation in such sources to substantially less than ~ 0.1% (though further significant noise reductions by pulse averaging are not feasible). Though marginally better than low repetition rate lasers, however, neither laser type exhibits a particularly high sensitivity for absorption spectroscopy by measuring directly the transmission through a sample as a function of wavelength. By contrast, dramatic improvements in sensitivity become possible in pump-probe spectroscopy by the use of mode-locked lasers in a way that is not possible with low repetition rate lasers. Since the essence of pump-probe spectroscopy is to compare the flux of photons transmitted through a sample when preceded by a perturbing pump pulse with the transmission in the absence of the pump pulse, it is possible to impose an intensity modulation on a train of pump pulses and to record the appearance of that modulation on the transmitted probe pulses following mixing of the two beams in the sample. The means whereby this is done are displayed diagrammatically in figure 2.

The pump and probe lasers of figure 2 are commonly derived from separately tuneable mode-locked lasers having identical pulse repetition rates and, as near as possible, identical beam profiles. One of the lasers is passed through an electro-optic modulator EOM/Polariser combination to yield an intensity modulation superimposed on the train of picosecond pulses, thereby becoming identified as the pump laser. The pulse train of the other laser, the probe, is left unperturbed, other than through the optical mixing process which occurs in the sample as a result of the varying transmission characteristics of the probe under the perturbing influence of the pump pulses (equations (10) and (11)). Because of the change in the sample transmission of those probe pulses following individual pump pulses, as compared with their transmission with no preceding pump pulse, a small fraction of the modulation impressed on the pump laser is transmitted to the probe. This small ac signal voltage is readily abstracted from the output of the photodiode detector (PD) by the lock-in-amplifier (L-I-A) referenced to the modulation frequency impressed on the pump by the electro-optic modulator (EOM) driven by an appropriate function generator. Dividing the rectified signal voltage abstracted by the phase-sensitive-detector by the d.c. signal voltage output of the detector provides a direct measure of the "gain" of equations (10) and (11).

Clearly in this type of spectroscopy it is essential to separate efficiently pump and probe beams post-sample to ensure that no element of the pump modulation is recorded by the detector, and hence processed by the phase sensitive detector. For different wavelength pump and probe lasers this is readily accomplished by means of diffraction gratings or wavelength selective filters. Such methods are, of course, not adequate where pump and probe beams derived from the same laser are of identical wavelengths. Under such circumstances, the most appropriate methods of separating pump and probe laser beams rely on the use of efficient polarisers for rejecting the linearly polarised probe beam from a pump beam which is set in orthogonal polarisation.

A steady displacement of the corner cube retro-reflector (CC) allows for a controlled time delay between individual pump and probe pulses from probe-before-pump (no signal voltage) to probe-after-pump, the magnitude of the signal voltage as a function of the pump-probe time delay then registering the rate of return of the system to the equilibrium condition, thereby providing information on the state lifetimes.

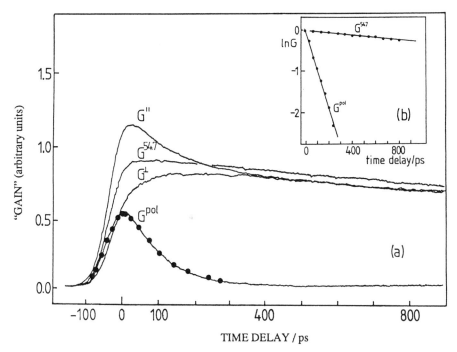

Figure 3. Gain decay curves for cresyl violet in solution in methanol. The various "gain" decay curves are plotted for different polarisation conditions of pump and probe beams. $G^{||}$ - parallel polarisations of pump and probe lasers (equation (12a)); $G^{\perp}$ - perpendicular polarisations (equation (12b)); $G^{54.7}$ - pump and probe polarisations inclined at 54.7^{0} (equation (12c)); G^{pol} - polarisation modulation of pump laser (equation (12d)). The $G^{54.7}$ and G^{pol} data of (a) are replotted in (b) in semi-log form.

A feature of the use of mode-locked laser systems in the above experiments is that the pulse repetition rate is so rapid that a fast amplitude (or polarisation) modulation may be superimposed on the pulse train by means of the electro-optic modulator. With a 80 M Hz pulse train from a mode-locked laser, for example, a 10 M Hz superimposed modulation yields a modulated train of pump pulses with four pulses "on" and four pulses "off" in each modulation cycle. The importance of this very fast modulation is that the noise spectrum in the output of these mode-locked laser systems diminishes by many orders of magnitude at 10 M Hz as compared with modulation frequencies below 100 K Hz. As a result, it becomes possible by using such rapid modulation methods and very fast phase-sensitive detectors to measure ac signal fluctuations in the probe laser power some $\sim 10^{8}$ times smaller than the average power. With this ability to measure to the shot-noise limit these small fractional changes induced by the pump pulses the potential is present for recording a wide variety of spectroscopic processes at an exceptionally high sensitivity limit.

PUMP-PROBE SPECTROSCOPY WITH PICOSECOND LASERS

Some of the scope of the high frequency modulation methods for studying state lifetimes is displayed clearly in figure 3 which shows temporal decays of cresyl violet in methanol solution for various polarisations of pump and probe lasers. In this study (Baran et al., 1984) the excited state of the cresyl violet absorption near 600 nm is pumped by 631 nm mode-locked pulses (modulated at 10 MHz), the concentration perturbation being probed by a similar train of probe pulses of wavelength 594 nm, the probe laser being located in the spectral region dominated by ground state absorption.

Since the absorbing chromophore in this instance is a large organic dye molecule with extensive conjugation, the absorption of a pump photon creates an excited electronic state with a charge distortion oriented along the long molecular axis. Absorption of significant numbers of linearly polarised pump photons by the sample thus creates an ordered array of excited state species and an ordered array of "holes" in the hitherto random distribution of molecular dipoles in the ground electronic state. Following pump excitation this perturbation relaxes towards the randomised equilibrium condition at a rate which depends on the excited state lifetime, τ_F, and the orientational relaxation time, τ_{or}. When the molecular ordering and state lifetime are taken into account, the probe laser "gain" decays with time according to the expression

$$G^{||} \propto (1 + \tfrac{4}{5} e^{-t/\tau_{or}}) e^{-t/\tau_F}, \tag{12a}$$

for parallel polarisations of pump and probe lasers (Baran et al., 1984),

and
$$G^{\perp} \propto (1 - \tfrac{2}{5} e^{-t/\tau_{or}}) e^{-t/\tau_F}, \tag{12b}$$

for orthogonally polarised lasers. The proportionality constant in these expressions is similar to the expression given in equation (10).

These temporal decay curves for parallel and perpendicular pump and probe polarisations are shown by curves $G^{||}$ and $G^{\perp}$ of figure 3. Although fitting of these curves by the theoretical expressions of equations (12a) and (12b) is perfectly feasible, a more convenient method of interpreting the perturbation created by the pump laser is to set the polarisations of probe and intensity-modulated pump lasers at the "magic angle", 54.7°, under which condition the decay depends only on the state lifetime

$$G^{54.7} \propto e^{-t/\tau_F}. \tag{12c}$$

Further, the use of a polarisation in place of an intensity modulation of the pump laser also yields a simple exponential decay

$$G^{pol} \propto e^{-t(1/\tau_F + 1/\tau_{or})}, \tag{12d}$$

from which τ_{or} can easily be determined once τ_F is measured. Figure 3 displays experimental examples of equations (12a) - (12d), analyses of which yield values of the orientational relaxation and state lifetimes of 107 ps and 3.6 ns, respectively (Baran et al., 1984). An indication of the sensitivity of the method is reflected in the signal-to-noise ratio in these traces for a solution concentration as low as $\sim 10^{-6}$ M.

Perhaps the most clear cut examples of the sensitivity of these concentration-modulation methods in the study of photoinduced absorption/bleaching, however, are found in the study of atoms. Figure 4 shows an example of the concentration-modulated absorption spectrum of excited states of neon near 134,000 cm^{-1} excited in a see-through optogalvanic discharge lamp. The atom concentrations in the four lowest excited states of neon (at 134,044 cm^{-1}, 134,461 cm^{-1}, 134,821 cm^{-1} and 135891 cm^{-1}) leading to the absorption features of figure 4 are so low (typically 10^{12} - 10^{10} cm^{-3}) that with the ~ 1 cm^{-1} line width of the picosecond lasers employed transitions from these levels in absorption are scarcely identifiable in the direct transmission spectrum of figure 4a.

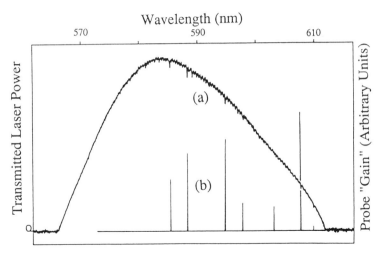

Figure 4. Comparison of the absorption and COMAS spectra of atomic neon excited by a current of 10 mA in a see-through hollow cathode optogalvanic lamp (Griffiths, 1990). For the direct transmission spectrum (a) the laser is passed 4 times through the 2 cm cathode: for the COMAS spectrum (b) the pump and probe lasers are focused together and passed singly through the lamp. In each case the picosecond Rhodamine 6G dye laser is scanned over its tuning range. The ordinate scales are arbitrary linear scales of the transmitted laser power and the probe "gain".

By contrast the COMAS spectra of figure 4b are observed with exceptionally high signal to noise ratios suggesting that atom concentrations in the range 10^5 - 10^6 atom cm^{-3} should readily be detectable using these methods.

Interestingly, apart from information on state lifetimes which may be obtained by delaying the probe in time following pump excitation, information is also available on the 3-dimensional distribution of these species in space. Figure 5 shows an example of such a study of excited state species concentrations in the hollow cathode neon discharge lamp. Integration of equation (7) over sections of the overlap regions of pump and probe beams in the vicinity of the focus shows that some 50% of the total signal is created in the confocal zone of dimension $2\pi w_0^2/\lambda$, and ~ 90% in a distance some three times greater than this. With a focal spot size of a few microns and a confocal parameter about an order of magnitude greater the potential clearly exists for measuring the species concentration in this localised zone, and therefore by moving this localised zone longitudinally and laterally through the discharge zone it proved possible to measure the distributions in 3-D space through the 4 cm x 2.5 mm anode-cathode region for all four excited state species of neon (Griffiths et al., 1990; Griffiths, 1990). Approximate distribution profiles of the lateral concentration gradients within the lamp are, of course, readily obtained by many methods, even by direct transmission measurements: longitudinal distribution profiles through the lamp of the sort shown in figure 5, however, are very much more difficult to attain other than via the non-linear spectroscopic techniques, of which COMAS is a particular example.

Ideally the longitudinal profiles should be determined with collinear pump and probe beams, the theoretical "gain" expression for which is given by equation (11). In practice, however, because of the length and the restrictive aperture of the lamp the spatial resolution of this approach, determined by a confocal parameter of ~ 2 mm, did not prove adequate to resolve the discharge structure. In order to obtain improved resolution pump and probe beams were separated by 4.3 mm at the focusing lens to provide a pair of beams intersecting at the focus, giving much better spatial resolution at a cost of a factor of ten

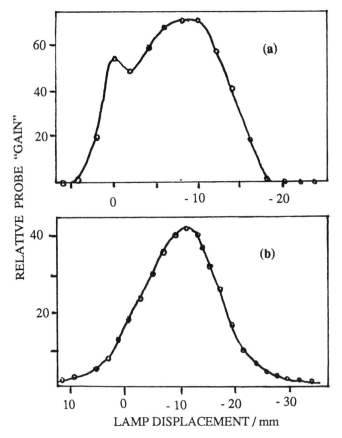

Figure 5. Longitudinal "gain" variations within a iron-neon "see-through" hollow cathode discharge lamp at a current of 5 mA (Griffiths et al., 1990). Pump and probe lasers at a wavelength of 609.8 nm excite the Ne transition $2p_4$ - $1s_4$. (a) - Crossed pump and probe beams; (b) - Collinear beams.

degradation in signal to noise and the loss of the quantitative expression for the probe "gain". The improved resolution in figure 5(a) serves to show the presence within the discharge of two maxima, the larger being positioned centrally within the lamp, the smaller in the region between the cathode and the co-axial anode.

The quantitative nature of concentration-modulated absorption spectroscopy was established in a study of lithium injected into a hydrocarbon flame (Langley et al., 1985, 1986), the direct absorption and "gain" spectra of the 2p ^{2}P - 2s ^{2}S transitions of lithium for various concentrations of the lithium salt solution aspirated into the flame being shown in figure 6. Since the absorbance values of figure 6 were measured directly under the same conditions as were used to measure the probe "gain", the relationship between them could be established from these data. Investigation of equation (11) shows that for the same pump and probe wavelengths (671 nm), the "gain" is expected to vary quadratically with the absorption cross section but only linearly with the species concentration. As is well known, the absorbance of a sample varies linearly with both the absorption cross-section and the species concentration. Because of this different dependence of the absorbance and the gain on σ and N_1^e it is possible to combine the absorbance and "gain" values to determine the species concentration directly with no further knowledge concerning the absorption cross section of the sample, thereby overcoming one of the major shortcomings in the use of the Beer-Lambert relationship, the inability to measure σ or N_1^e directly unless

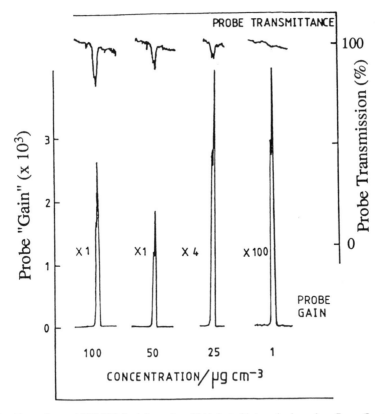

Figure 6. Absorption and COMAS "gain" spectra of Li injected into a hydrocarbon flame (Langley et al., 1985, 1986). The spectra were obtained with a picosecond DCM laser at 671 nm. The right hand scale refers to the absorption spectrum, the left hand scale to the "gain" spectrum.

the other parameter is known. This relationship was clearly established in the study by the straight line relationship observed between the square root of the "gain" and the sample absorbance (Langley et al., 1985, 1986). From this relationship it was possible to measure directly the Li atom concentration in the flame for each solution with no further knowledge concerning the absorption cross section of the sample. Estimates of the sensitivity limit for the detection of this transition derived from the signal to noise ratios recorded showed that solution concentrations as low as 10^{-4} - 10^{-5} ppm should readily be measurable, an improvement over present sensitivity limits for Li atom detection of some 3-4 orders of magnitude.

Confirmation of this sensitivity is shown in figure 7 which displays not only the temporal decay of the "gain" for the 2p ^{2}P - 2s ^{2}S transition in absorption from the ^{2}S Ground State (figure 7(a)), but also the very much weaker related 3d ^{2}D - 2p ^{2}P transition of lithium at 611 nm in the same flame (figure 7(b)). These traces show the variation of the probe signal voltage as the probe laser is delayed in time following pump excitation. The lower electronic state involved in the trace of figure 7(b) is the 2p ^{2}P level at ~ 14,910 cm^{-1}, the fractional population of this level at the flame temperature being less than 10^{-4} of the ground state 2s ^{2}S population. Notwithstanding the very low atom concentration in the focal zone of the lasers (~ 10^8 2p ^{2}P Li atoms per cubic centimetre in the flame) the signal-to-noise ratio is sufficiently good that a well defined state lifetime could be measured (Beaman et al., 1986). This is not a simple exponential decay process since there are many kinetic processes involved, including the deactivation of the excited state 3d ^{2}D level as well as the decay of 2p ^{2}P to the 2s ^{2}S level. Combination of the study of the 2p ^{2}P - 2s ^{2}S transition with the results from the decay of 3d ^{2}D - 2p ^{2}P yielded lifetimes of 216 ± 20 ps and 528 ± 20 ps for 3d ^{2}D and 2p ^{2}P respectively (Beaman et al., 1986), results which are dramatically shorter than the values derived for these same states

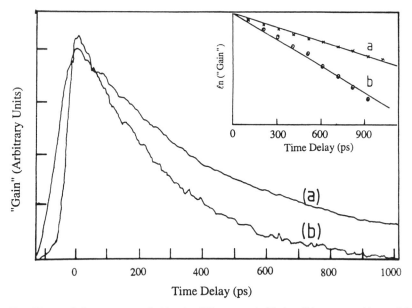

Figure 7. Temporal decay curves of: (a) the 671 nm: and (b) the 611 nm transitions of Li in a hydrocarbon flame (Beaman et al., 1986): (a) depends only on the lifetime in the 2p state and gives a simple exponential decay, (b) depends on the lifetimes of both the 2p and the 3d states, which gives rise to more complex decay kinetics. The inset diagram displays part of the same data displayed in semi-log format.

in a low pressure discharge environment because of the marked influence of frequent deactivating collisions occurring in the flame environment.

The methods described above are not the only ways of studying spectroscopic events occurring on the picosecond time scale, although they do represent probably the most sensitive methods currently available for recording such processes. Further, they have the advantage that they are quantitative and do not require the presence of a fluorescing entity to be applicable. They are completely general yielding information on absolute species concentrations, electronic state lifetimes and orientational relaxation times to the picosecond domain. The only limitation would appear to arise from the need for a tuneable picosecond laser source capable of perturbing state populations significantly (usually a few milliwatts of mode-locked power is quite adequate), so that the potential exists for direct measurement of absolute species concentrations for any entity possessing optical transitions in the visible/ultraviolet regions of the electromagnetic spectrum.

ACKNOWLEDGEMENTS

I am indebted to the many co-workers and postgraduate students whose work is described here, and to the Engineering and Physical Sciences Research Council for the award of research grants, with the aid of which the work was carried out.

REFERENCES

Baran, J., Langley, A.J., and Jones, W. J., 1984, *Chem. Phys.* 87:305.

Beaman, R.A., Davies, A.N., Langley, A.J., Jones, W. J., and Baran, J., 1986, *Chem. Phys.* 101:127.

Bradley, D.J., 1977, Methods of Generation, *in*: Topics in Applied Physics, Vol. 18, "Ultrashort Light Pulses," S. L. Shapiro, ed., Springer-Verlag, Berlin.

Demtröder, W., 1982, Springer Series in Chemical Physics 5, "Laser Spectroscopy: Basic Concepts and Instrumentation," Springer-Verlag, Berlin.

Fleming, G.R., 1986, "Chemical Applications of Ultrafast Spectroscopy," Oxford University Press, Oxford.

Griffiths, T.R., 1990, Ph.D. Thesis, University of Wales.

Griffiths, T.R., Jones, W.J., and Smith, G, 1990, Concentration-Profiling in a Neon Optogalvanic Lamp Using Concentration-Modulated Absorption Spectroscopy, *in*: "Optogalvanic Spectroscopy," R.S. Stewart and J.E. Lawler, eds., Inst. Phys. Conf. Ser. No. 113(7), IOP Publishing (1991).

Ippen, E.P., and Shank, C.V., 1977, Techniques of Measurement, *in*: Topics in Applied Physics, Vol. 18, "Utrashort Light Pulses," S. L. Shapiro, ed., Springer-Verlag, Berlin.

Langley, A.J., Beaman, R.A., Baran, J., Davies, A.N., and Jones, W. J., 1985, *Opt. Letters* 10:327.

Langley, A.J., Beaman, R.A., Davies, A.N., Jones, W. J., and Baran, J., 1986, *Chem. Phys.* 101:117.

Mallawaarachchi, W., Davies, A.N., Beaman, R.A., Langley, A.J., and Jones, W. J., 1987, *J. Chem. Soc., Faraday Trans. 2* 83:707.

Czarnik-Matusewicz, B., Griffiths, R., Jones, P.F., Jones, W.J., Mallawaarachchi, W., and Smith, G., 1988, *J. Chem. Soc., Faraday Trans. 2* 84:1867.

Shapiro, S.L., ed, 1977, Topics in Applied Physics, Vol. 18, "Utrashort Light Pulses," S. L. Shapiro, ed., Springer-Verlag, Berlin.

Yariv, A., 1985, "Optical Electronics," 3rd Edn., Holt-Saunders, New York.

RAMAN SPECTROSCOPY

Daniel Wolverson

School of Physics
University of East Anglia
Norwich, NR4 7TJ, UK.

INTRODUCTION

Raman scattering is the inelastic scattering of light by a material; the word "inelastic" implies that energy is transferred between the light quanta and the material, so that the scattered light may have a longer or shorter wavelength than the incident light. The study of the spectrum of the light scattered from a particular material is therefore termed Raman spectroscopy and is of interest because, as will be seen, information can be gained about the structure, the composition and the vibrational or electronic states of the scattering material. Raman spectroscopy is a large field, with many variations on the basic technique and with many new applications being found each year. It is not possible here to review all the areas of industrial and academic research in which Raman spectroscopy plays a role, so that, after a brief introduction to the theory of the topic, some examples of the applications of Raman spectroscopy will be discussed. Some illustrations of recent developments in experimental methods will also be given. For further examples of industrial applications the reader is referred to the chapter by Everall.

ELEMENTARY THEORY OF RAMAN SCATTERING

The purpose of this section is to give an insight into how Raman scattering arises and how the selection rules that govern it can be understood. The theory of Raman scattering cannot really be described as elementary, since both group theory and high-order perturbation theory are required for a proper discussion of the phenomenon. The interested reader is referred to the many excellent textbooks on the subject, for example, Anderson (1971), and similar treatments are found in many other books. For thorough treatments of the quantum mechanical theory see, for example, Loudon (1978), or the series of texts edited by Cardona and Güntherodt (1983).

An Introduction to Laser Spectroscopy
Edited by David L. Andrews and Andrey A. Demidov, Plenum Press, New York, 1995

The Classical Theory

The interaction of light with a material can be thought of in a classical picture as being due to the action of the electric field of the light wave on the charges in the material. For simplicity, the following sections consider the interaction of light with a single molecule, though the results will be quite general. This discussion is based on that given by Demtröder (1988).

If P_i represents the dipole moment (a vector) of a molecule and E_i is the electric field vector of the light, then a general expression for the dipole moment is given by (1), in which the subscripts i, j, k, l... each run over the three spatial directions x, y, z:

$$P_i = p_i + \alpha_{ij} E_j + \beta_{ijk} E_j E_k + \gamma_{ijkl} E_j E_k E_l + ... \tag{1}$$

This expression includes the possibility of a permanent dipole moment p_i as well as several components induced by the electric field of the light. Of these, the most important here is the second term, which is linear in E; α_{ij} is referred to as the polarisability and is, in general, a tensor. The terms in higher powers of the incident electric field give rise to non-linear processes such as hyper-Raman scattering; these will not be discussed further here. Both the permanent dipole of the molecule (if any) and the polarisability may change if the molecule vibrates. One can expand p_i and α_{ij} as Taylor series in terms of generalised co-ordinates q_n which describe the n vibration normal modes:

$$p_i = p_{oi} + \sum_n \left(\frac{\partial p_i}{\partial q_n} \right)_{q=0} q_n$$

$$\alpha_{ij} = \alpha_{oij} + \sum_n \left(\frac{\partial \alpha_{ij}}{\partial q_n} \right)_{q=0} q_n \tag{2}$$

If the atomic displacements are assumed to be small, one can approximate the time-dependence of the atomic displacements q_n (of frequency ω_n) and of the electric field of the light (of frequency ω_L) in the following way:

$$q_n(t) = q_{0n} \cos(\omega_n t)$$

$$E_i(t) = E_{0i} \cos(\omega_L t) \tag{3}$$

These expressions can then be substituted into a version of (1) that has been simplified by neglecting the non-linear terms (β, γ, etc.) and the result can be expanded:

$$P_i = p_i + \alpha_{ij} E_j + ...$$

$$= p_{oi} + \sum_n \left(\frac{\partial p_i}{\partial q_n} \right)_{q=0} q_{0n} \cos(\omega_n t) \quad + \quad \alpha_{oij} E_{0j} \cos(\omega_L t)$$

$$+ \sum_n \left(\frac{\partial \alpha_{ij}}{\partial q_n}\right)_{q=0} q_{0n} \cos(\omega_n t) \times E_{0j} \cos(\omega_L t) + \ldots$$

$$= p_{oi} + \sum_n \left(\frac{\partial p_i}{\partial q_n}\right)_{q=0} q_{0n} \cos(\omega_n t) \quad + \quad \alpha_{oij} E_{0j} \cos(\omega_L t)$$

$$+ q_{0n} E_{0j} \sum_n \left(\frac{\partial \alpha_{ij}}{\partial q_n}\right)_{q=0} \{\cos(\omega_L + \omega_n)t + \cos(\omega_L - \omega_n)t\} + \ldots$$

$$\text{(4)}$$

The physical meaning of each term in equation (4) must now be clarified; to do this, each of the important terms of the right hand side of the equation are listed and discussed in turn. The first term of interest is as follows:

$$\sum_n \left(\frac{\partial p_i}{\partial q_n}\right)_{q=0} q_{0n} \cos(\omega_n t) \quad . \tag{5}$$

This term represents the sum of the periodic changes in the dipole moment of the molecule which may be induced by each of the n molecular vibrations; the frequency of the modulation of the dipole moment by a given vibrational mode is of course the vibrational frequency ω_n and, as long as $(\partial p_i / \partial q_n)_{q=0}$ is non-zero, the molecule can exchange energy with light of a frequency equal to the vibrational frequency. Thus, infra-red (IR) absorption is possible and such vibrational modes are termed "IR-active" modes. As will be discussed in more detail below, whether or not a particular mode has a non-zero derivative term depends on the nature of the atomic displacements involved in that mode.

The next term is :

$$\alpha_{oij} E_{0j} \cos(\omega_L t) \quad . \tag{6}$$

This represents the fact that the incident light induces a dipole moment which oscillates at the same frequency as the light, so that light of that frequency is re-radiated: this is elastic or Rayleigh scattering. The induced dipole moment is proportional to the polarisability, which may be modulated by the molecular vibration. Equation (6) arises from the constant part of the polarisability and the final term, (7), contains the oscillating part:

$$q_{0n} E_{0j} \sum_n \left(\frac{\partial \alpha_{ij}}{\partial q_n}\right)_{q=0} \cos(\omega_L \pm \omega_n)t \quad . \tag{7}$$

This term represents the part of the dipole moment which is induced by the light and is also proportional to the oscillating part of the polarisability. The time-dependence of this term reflects the product of the frequencies of the incident light and of the vibrational mode. Thus, if $(\partial \alpha_{ij} / \partial q_n)_{q=0}$ is non-zero, two "beat" terms arise corresponding to light re-radiated at the sum and difference frequencies; this is Raman scattering (and the vibrational mode would be termed "Raman-active"). Conventionally, the scattering of light with a decrease in frequency is referred to as Stokes scattering and the converse is called anti-Stokes scattering.

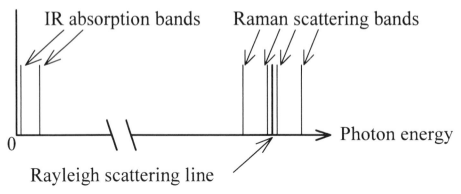

Figure 1. Schematic diagram of the absorption and emission transitions predicted by the classical theory outlined in the text; the vertical scale is arbitrary.

Figure 1 illustrates the different processes discussed above for a hypothetical system with two vibrational modes (which are assumed to be both Raman-active and infra-red active). The absorption bands appear at photon energies equal to the energies of the vibrational quanta (on the left of the figure), the Rayleigh scattering appears at the photon energy of the incident light and the Stokes and anti-Stokes Raman bands appear near the photon energy of the incident light (on the right of the figure).

The Quantum Mechanical Theory

An experimental Raman spectrum (of CCl_4) is shown in figure 2 (Gardiner and Graves, 1989). It is clear that the Stokes scattering, which involves loss of energy from the light to the molecule, gives a stronger signal than anti-Stokes scattering. This is in disagreement with equation (7) (which predicts equal intensities for both processes) and is the most obvious indication of the breakdown of the classical picture.

The quantum-mechanical theory of Raman scattering solves this problem. In the quantum picture, the molecular vibrations are quantised. The scattering process is viewed as the creation and annihilation of vibrational excitations (or "phonons") by photons. This interpretation leads naturally to the fact that photons can always create vibrational quanta in a system, so that the probability of Stokes scattering is not temperature-dependent. However, the probability of *annihilation* of a vibrational quantum depends on the probability of finding the system in an excited vibrational state. The ratio of the intensities of Stokes and anti-Stokes bands thus reflects the Boltzmann factor $\exp(-h\nu/kT)$ where $h\nu$ is the vibrational quantum energy (but note that the intensity ratio does not depend *only* this factor: see Gardiner and Graves (1989)). The following diagram, figure 3, illustrates the transitions involved in Raman scattering; here, n is the number of vibrational quanta present. The separation in energy of the vibrational levels is assumed to be much smaller than that of the electronic states.

Excitation by the incoming light raises the system to an excited electronic state, from which return to a different vibrational level of the electronic ground state may be possible. Stokes scattering involves the loss of energy from the incident photon to the material and thus the return of the system to a state with a higher vibrational quantum number.

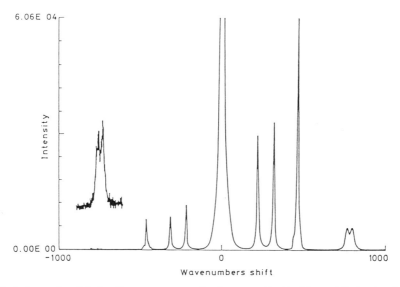

Figure 2. Stokes and anti-Stokes Raman spectrum of carbon tetrachloride. The anti-Stokes band at about 790 cm^{-1} has been magnified by a factor of 100 (Gardiner and Graves, 1989)

The nature of the excited electronic state deserves some comment. In general, Raman scattering should not be thought of as a purely sequential process of light absorption at one wavelength followed by re-emission of light at a different wavelength, as might happen if the intermediate state were an energy eigenstate of the system. Rather, the energy of the system is increased by an amount equal to the energy of an incident photon for a very brief time determined by the uncertainty principle. In the language of quantum mechanics, the system is said to be in a "virtual" excited state.

It is of course possible to adjust either the incident or emitted photon energies so that one of them *does* correspond to an electronic transition energy. It is then often observed that the Raman scattering cross-section is dramatically increased, an effect which is known as resonant Raman scattering. This phenomenon would occupy too much space to describe properly here and good texts are available (for example, Cardona and Güntherodt, 1983).

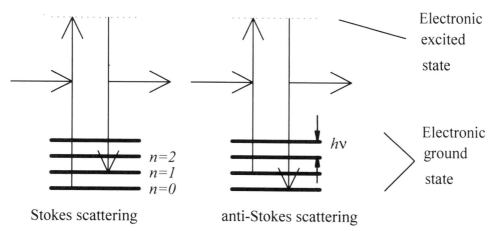

Figure 3. Schematic diagram (not to scale) of the Raman scattering process (the vertical direction represents energy and hν represents the vibrational quantum of energy).

Resonant enhancement of the scattering cross-section is often exploited experimentally to obtain stronger signals or even to investigate the electronic energy levels of the system by measuring the strength of Raman signals as a function of the excitation wavelength (an example of this is given below).

SELECTION RULES

Where the vibrational spectroscopy of simple molecular systems is concerned, the selection rules that govern Raman scattering constitute one of the main features distinguishing it from IR absorption. The Raman selection rules can be understood by reference to equation (7), which is re-written below in a simplified form; E_{si} is the electric field vector of the scattered light.

$$E_{si} \quad \propto \quad P_i \quad \propto \quad \left(\frac{\partial \alpha_{ij}}{\partial q_n}\right) E_{0j} . \tag{8}$$

This equation shows that the incoming and outgoing electric fields (E_{0j} and E_{si}) are related, for a given vibrational mode n, by the Raman tensor $\mathbf{a_n}$ defined in (9); if the element with subscripts i, j is non-zero, then Raman scattering is possible between E_{0j} and E_{si}. We shall consider this important point in more detail. First, it is helpful to write out explicitly the tensor equation relating the incoming and outgoing light waves:

$$\begin{pmatrix} E_{sx} \\ E_{sy} \\ E_{sz} \end{pmatrix} = \begin{pmatrix} a_{xx} & a_{xy} & a_{xz} \\ a_{yx} & a_{yy} & a_{yz} \\ a_{zx} & a_{zy} & a_{zz} \end{pmatrix}_n \begin{pmatrix} E_{ox} \\ E_{oy} \\ E_{oz} \end{pmatrix} . \tag{9}$$

Group theory (beyond the scope of this discussion) allows one to determine the symmetry species of the vibrational modes and also which elements of the Raman tensor $\mathbf{a_n}$ are non-zero. Raman tensors are tabulated in many books and their use explained (for example, see

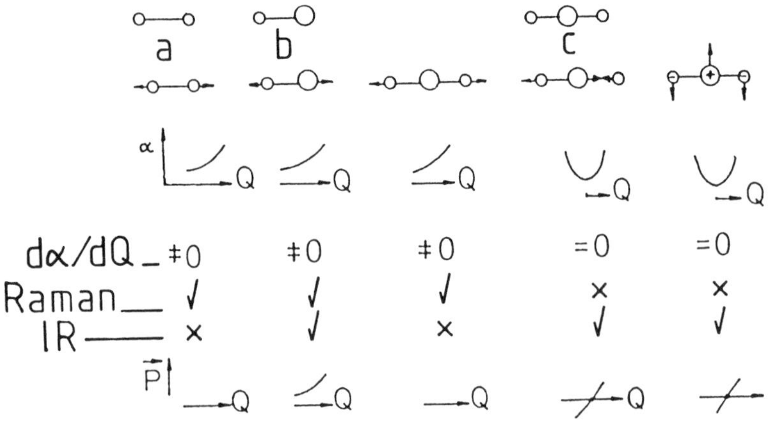

Figure 4. The selection rules for infra-red absorption and Raman scattering illustrated for three simple molecules: ✓ indicates that a mode is active and ✗ that it is not (with permission from Kuzmany, 1989).

Kuzmany, 1989, in German; or Boardman, 1973). In order that this discussion is not too abstract, we shall now consider in elementary terms how different vibrational modes of simple molecules may be IR-active or Raman-active.

Figure 4 demonstrates how one can visualise the origins of the selection rules. Three simple molecules are considered, shown in the first row of the diagram. The second row indicates the vibrational modes under consideration. In the third row, the change of polarizability α for a generalised atomic displacement Q is indicated; from this, it is easy to see whether a given mode has a non-zero value of the derivative term (equation 8). Likewise, in the bottom row, the effect of the atomic displacements on the polarizability of each of the molecules is sketched, allowing one to conclude whether that mode is infra-red active (by consideration of the derivative term in equation 5).

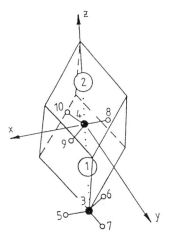

Figure 5. The unit cell of calcite, $CaCO_3$. ① and ② are Ca^{2+} ions; 4, 8, 9, and 10 form one CO_3^{2-} ion (with permission from Porto et al., 1966).

For example, for the first vibrational mode of molecule b, movement of the atoms towards or away from each other certainly changes the dipole moment and may well change the polarisability; therefore, the mode may be detected in both Raman scattering and IR absorption experiments. Detailed quantum mechanical calculations are needed if one wishes to calculate *how much* the polarisability is modulated by the vibration and therefore how strong the Raman scattering is, but this information is often not required.

To see how Raman spectra can vary in different experimental geometries, we follow Kuzmany (1989) in considering the example of calcite (from a classic early work in the field of Raman spectroscopy by Porto et al.,1966). Figure 5 shows the calcite unit cell, which contains two Ca^{2+} ions and two CO_3^{2-} ions. The arrangement of these ions is such that there is a high-symmetry axis which is conventionally labelled z, as in the figure. The direction in which the orthogonal x and y axes are defined is arbitrary. Without detailed group theoretical analysis, we will simply state that the vibrations of the unit cell can in this case be classified into various symmetry types including two with conventional labels A_{1g} and E_g. The Raman tensors for these have the following form:

$$A_{1g}: \begin{pmatrix} a & 0 & 0 \\ 0 & a & 0 \\ 0 & 0 & b \end{pmatrix} \qquad E_g: \begin{pmatrix} c & 0 & d \\ 0 & c & 0 \\ d & 0 & 0 \end{pmatrix} \text{ or } \begin{pmatrix} c & 0 & 0 \\ 0 & -c & d \\ 0 & d & 0 \end{pmatrix}. \qquad (10)$$

By referring to equation (9), we can expect Raman scattering from light polarised along the x axis to yield light polarised along the x axis for A_{1g} vibrational modes and along x (relative intensity c^2) or z (relative intensity d^2) for E_g vibrational modes. On the other hand, if the incident light is polarised parallel to the high-symmetry z axis, Raman scattering from an E_g mode will be polarised only along the x axis whilst Raman scattering from an A_{1g} mode will be polarised only along the z axis. This is borne out by experiment, as shown in figure 6.

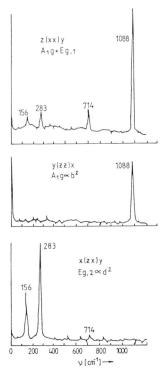

Figure 6. The Raman spectra of calcite in three different scattering geometries, Porto et al. (1966).

Figure 6 also illustrates the Porto notation, a useful convention for representing experimental scattering geometries. The symbol x(zx)y, for example, in the third spectrum of figure 6 indicates that the excitation light was incident on the sample along the x axis and was polarised along the z direction, whilst the light that was detected was travelling along y and was polarised along x. This notation is most useful if the axes are defined with respect to the crystal axes of symmetry (as here) so that x...z relate directly to components of the Raman tensor, rather than using arbitrary "laboratory" axes.

EXAMPLES OF WHAT CAN BE LEARNT

This section gives some diverse examples of applications of Raman spectroscopy in contexts where other forms of spectroscopy would not be appropriate. As has been described, Raman spectroscopy can be used to obtain information about the vibrational spectrum of a material. This can lead to a better understanding of the chemical composition, especially in many organic compounds where different chemical groups have very characteristic vibrational frequencies. It may also be possible to deduce the structure of a material; for example, in molecular systems, the juxtaposition of different chemical species leads to small modifications of the usual vibrational frequencies, so that longer-range structure may be investigated. In crystalline solids, the selection rules for scattering from the lattice vibrations (phonons) indicate the symmetry of the crystal unit cell and can thus reveal (for example) phase changes. Finally, via resonance Raman spectra, one can sometimes correlate a particular vibrational mode with a particular electronic transition energy and thus learn more about the electronic structure of the material. The interests of the Raman spectroscopy group at the University of East Anglia are, of course, reflected in the examples which follow.

Example 1: Composition and Structure of Phosphorus-Selenium Glasses

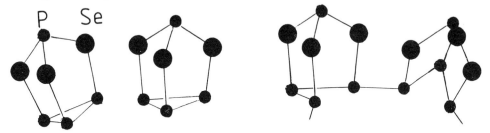

Figure 7. Two possible atomic configurations in P-Se glasses; on the left, discrete molecular P_4Se_3 cages are shown, whilst on the right, a continuous random network is indicated.

Figure 7 shows a sketch of "cage" molecules of P_4Se_3: in stoichiometric crystalline P_4Se_3, the vibrational modes involving deformations of single cages are clearly detectable and can easily be recognised, being at much higher energies (or, equivalently, Raman shifts) than the vibrational modes of the lattice in which whole groups of cages move together as rigid entities. However, it is possible to make non-crystalline materials (glasses) which contain

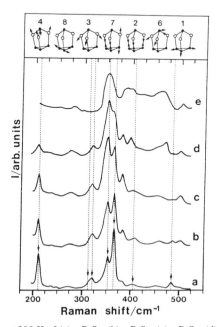

Figure 8. Raman spectra at 300 K of (a) a-P_2Se, (b) a-P_3Se, (c) a-P_4Se, (d) a-$P_{0.84}Se_{0.16}$ and (e) a-P

excess phosphorus. The question that then arises is whether the bonding within the cages is disrupted so that they "polymerise" into an infinite network, or whether the excess P (and perhaps some selenium) simply hinders the crystallisation of the solid, forming a matrix in which the cages remain intact. Figure 8 shows Raman spectra of a range of glasses, the nearest in composition to P_4Se_3 being P_2Se. For this glass, all the Raman-active vibrational modes of P_4Se_3 cages can be seen (figure 8a). As more P is introduced, these bands weaken whilst modes corresponding to those of amorphous P increase, until only the modes corresponding to the vibrations of amorphous P and of $P-Se_3$ pyramids remain (figure 8d). This shows that there is a progressive increase in the cross-linking of cages as the P content increases. This example is discussed more fully by Phillips et al. (1989).

Example 2: Phase Changes in Ammonium Nitrate

Ammonium nitrate is a material with a very rich phase diagram, having several different crystalline structures at different temperatures and pressures. Even more intriguing is the fact that it can show some "memory" effects, where it can remain in a non-equilibrium phase depending on its previous heat treatment. Differential scanning calorimetry (DSC) is therefore useful to investigate its phase changes. DSC can measure heats of melting, crystallisation, or other phases changes, but gives no direct information about the crystal structure; in order to help understand this complicated system, Raman spectroscopy has been combined with DSC to correlate thermal and vibrational information. Here, there is no change of composition; the interest lies in the re-arrangement of the same chemical species to give a new crystal structure. Figure 9 shows Raman spectra of NH_4NO_3 over a range of temperatures; a phase change is clearly visible. Note that in the central spectrum, recorded at the temperature of the phase change, Raman bands are observed which are quite unlike those of either the initial or final phases; it therefore appears that the phase change may proceed via an intermediate phase. For further details, see Harju et al. (1992).

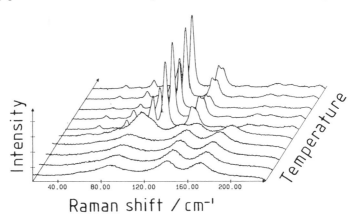

Figure 9.: Raman spectra of NH_4NO_3 undergoing the transition from phase IV to phase V.

Example 3: Structure of Carbon Films

A very active area of applied research at present is in the preparation of carbon films and coatings. Depending on the deposition conditions, one can coat materials in films composed of amorphous carbon, which may have useful electrical properties, or even "diamond-like" or polycrystalline diamond films, which form chemically-inert, hard-wearing surfaces. In figure 10, the Raman spectrum of such a film is shown. The vibrational bands of the silicon substrate, of some microscopic diamond crystals, and of a large amount of amorphous carbon can be seen; this spectrum indicates that in this particular case, the preparation technique has yet to be optimised.

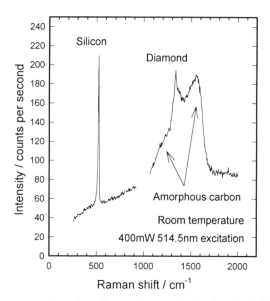

Figure 10. Raman spectrum of a carbon film on silicon, showing bands of diamond and amorphous carbon.

Example 4: Composition of Zinc Telluride Epitaxial Layers

Artificial multilayered semiconductor structures are extremely important in optoelectronics technology; for example, solid state laser diodes are constructed in this way. Consequently, there is much interest in the growth of new layered systems, of which

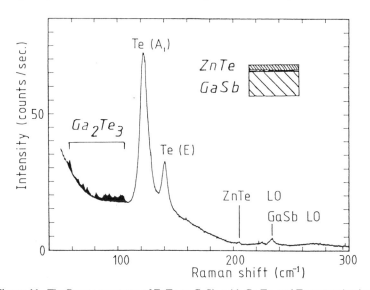

Figure 11. The Raman spectrum of ZnTe on GaSb, with Ga_2Te_3 and Te contamination

ZnTe upon GaSb is one example. Both these semiconductors have the zinc-blende crystal structure, in which one observes a vibrational spectrum consisting (depending on the experimental geometry) of one lattice mode, the longitudinal optic (LO) phonon. Figure 11 shows a Raman spectrum of a very thin layer of ZnTe on GaSb; the LO phonons of both materials can be seen, but are weak in this spectrum because the laser photon energy was chosen to be well below the electronic absorption edge of the layers and hence no resonant enhancement is observed. At this low photon energy the light, which is incident on the sample from above as drawn in the insert diagram, penetrates through the ZnTe layer with little absorption, and Raman scattering from the region of the interface between the two layers can be observed. Two very strong bands can be seen arising from pure Te; there are also several weaker bands at smaller Raman shifts which have been identified (by comparison with bulk material) as being due to Ga_2Te_3. The presence of these materials worsens the quality of the ZnTe layer and was eliminated in later structures by modifying the growth conditions. Further details are given by Halsall et al. (1992).

Example 5: Electronic States of Impurities in Semiconductors

Control of electronic conduction in semiconductors is achieved by doping, that is, by the introduction of small quantities of impurities whose valence is lower or higher than that of the elements whose crystal lattice sites they occupy. The dopants can, therefore, introduce free charge carriers into the system. In any prospective semiconductor material, then, an important question is to identify suitable elements to act as dopants. The present

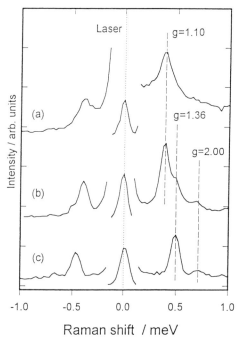

Figure 12. Spin-flip Raman spectra of ZnSe layers with different impurity concentrations, at 1.6 K and in a magnetic field of 5 Tesla. The laser line is shown in the centre of the figure, greatly attenuated. Negative values of the Raman shift (the left-hand side of the figure) here imply anti-Stokes scattering.

case concerns ZnSe, a material of interest for the production of blue light-emitting diodes and lasers. Chemical analysis of the layers shows that, although suitable amounts of nitrogen can be introduced (which do act as dopants), when a certain concentration is exceeded, further addition of nitrogen ceases to increase the doping level. It thus appears that high nitrogen concentrations may generate defects that capture the free carriers introduced by the nitrogen. Raman spectroscopy was used to investigate this problem, but rather than study vibrational spectra, the electronic modes of the impurities were probed directly.

Figure 12 shows Raman spectra of ZnSe layers with different nitrogen concentrations, at low temperatures and in a high magnetic field (5 Tesla). In high magnetic fields, the energy states of unpaired electrons undergo a Zeeman splitting; Raman scattering is then possible between the two spin states (so-called spin-flip Raman scattering). The splitting in energy of the spin states depends linearly on the magnetic field with a constant of proportionality (the effective gyromagnetic ratio, g) which is different for different defect centres and which can therefore help to distinguish between them. Here, the Raman scattering is thus comparable (to a certain extent) to electron spin resonance experiments.

Figure 12 also shows the laser line, attenuated and recorded in order to obtain accurate values of the Raman shifts of the spin-flip bands. The spectra show (from top to bottom) that as the nitrogen content increases, another defect in the crystal lattice is also introduced which gives rise to the signal with g ~1.36; this is the defect which counteracts the desired effects of the N impurity. Another signal (at g ~ 2.0) also appears and is related to a second type of defect centre. Note the scale of the Raman shifts on the figure; this experiment requires an unusually high resolution and very good rejection of the stray laser light, such as is obtained by use of a triple spectrometer (considered later). These results are discussed further by Boyce et al. (1993).

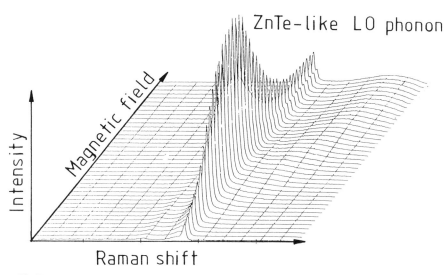

Figure 13. Raman spectra of $Zn_{0.96}Mn_{0.04}Te$ at 1.6 K from 0 to 3 Tesla (front to back respectively). The LO phonon Raman band at the centre of the figure has a Raman shift of about 215 cm^{-1}.

Example 6: Electronic States of Semiconductors via Resonance Raman

A good example of resonant Raman scattering is provided by the data shown in figures 13 and 14. Here, the materials concerned are semiconductors (ZnTe and CdTe) which contain manganese ions; these have unpaired 3d electrons that are not involved in chemical bonding, but which give the Mn^{2+} ion a permanent magnetic moment. The interaction of the valence electrons in the crystal with these localised magnetic moments means that the main absorption transition (the band gap) of the semiconductor is very sensitive to magnetic fields at low temperatures. In figure 13, we show the Raman scattering from the strongest lattice mode (a longitudinal optic, or LO, phonon) of the $Zn_{0.96}Mn_{0.04}Te$ recorded at a series of magnetic field strengths. As the magnetic field is increased from zero, the electronic transition energy of the sample is "tuned" into resonance with the laser energy, and the signal strength increases dramatically. Beyond magnetic fields of about 2 Tesla, the LO phonon signal decreases again as the resonance condition is passed.

A more usual form of resonance Raman spectroscopy involves tuning the laser energy to match an electronic transition of the sample. Figure 14 demonstrates this, showing not the Raman spectra themselves (which in this example are of spin-flip Raman scattering), but the integrated intensity of two Raman bands of a semiconductor multilayer as a function of the laser energy (such a spectrum is usually called a "resonance profile"). This spectrum is equivalent in some ways to an absorption spectrum, but is more selective, since the resonance energies observed can be correlated with the excitations that are enhanced. In figure 14, then, the two resonance profiles are obtained from different Raman bands arising from excitations of different layers of the sample. There are numerous instances in the literature of selective, resonant excitation of vibrational modes of chromophores attached to specific parts of complex organic molecules. The experimental technique developed to obtain these spectra is described in more detail by Railson et al. (1993), and the phenomenon of spin-flip Raman scattering is discussed by Halsall et al. (1994) and references therein.

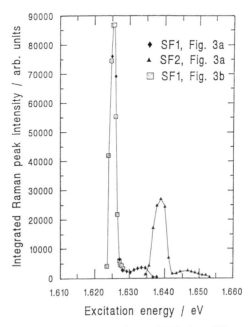

Figure 14. Resonance profiles of two Raman bands, SF1 and SF2, from different layers of a semiconductor multilayer sample, showing significant resonant enhancement over a very narrow energy range.

EXAMPLES OF EXPERIMENTAL TECHNIQUES

Only a brief discussion of techniques is possible here as the field has become immense; one useful book containing a great deal of practical information is that by Gardiner and Graves (1989). The traditional (and still widely used) experimental set-up for Raman spectroscopy is shown in figure 15.

Types of Laser

The different types of laser and their use as light sources for spectroscopy are the subject of other chapters of this book and will not be discussed here. Conventional Raman spectroscopy uses continuous wave lasers, many of which are described by Hecht (1986).

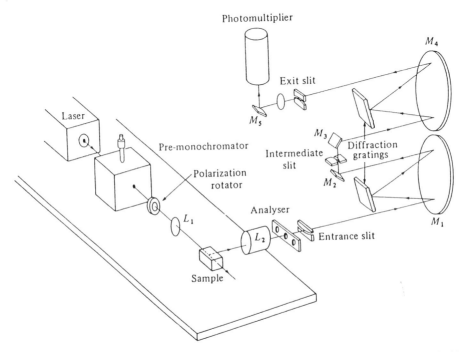

Figure 15. The layout of a traditional Raman spectroscopy system using a double spectrometer and with a photomultiplier as detector, with permission from Bower and Maddams (1989). Mirrors are labelled M.

Spontaneous Emission from the Laser

One comment on the use of lasers in Raman spectroscopy is that although lasers are often regarded as ideal monochromatic light sources, this is far from the case in practice. In the case of tunable lasers, there is always broad-band spontaneous emission over the whole tuning range of the gain medium, whilst in the case of ion lasers, there are many atomic transitions which give rise to sharp spectral lines; both of these can easily be comparable in intensity to the required signals. There are two common solutions; the most effective is to use a small monochromator (often a double-pass prism spectrometer, which is capable of handling large powers) which acts as a filter for the laser beam before it reaches the sample; such a filter is illustrated in figure 15. About 50% of the laser power is generally lost in this process. Another solution is to pass the beam through an aperture (or spatial filter). Since the spontaneous emission is radiated isotropically, much of it can be blocked before it is focused onto the sample along with the laser beam.

Choice of Laser Wavelength

The massive enhancement of Raman scattering when the excitation is resonant with some electronic transition of the material has already been mentioned. To exploit this effect, one may require a range of laser lines or even a tunable laser. However, other factors influence the choice of excitation wavelength. Many materials show strong photoluminescence (fluorescence) when the excitation lies close in energy to an absorption band. This may mean that it is advantageous to *avoid* resonance conditions. An example of this problem is shown in figure 16. One instrument which is popular for avoiding fluorescence problems is the Fourier Transform spectrometer, which typically uses a Nd:YAG laser giving a very low photon energy (wavelength 1064 nm) that is often insufficient to stimulate fluorescence.

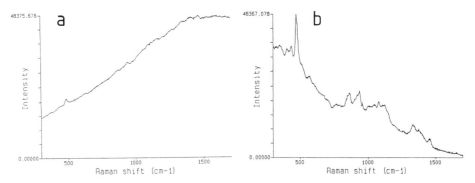

Figure 16. Raman spectra of rice flour at excitation wavelengths of (a) 514.5 nm and (b) 752 nm; at 514.5 nm, strong fluorescence is emitted by the sample which makes the weak Raman signals almost undetectable. Very little fluorescence is emitted for 752 nm excitation (courtesy of Spex Industries Inc).

Spectrometers

Many developments have been made recently in the area of spectrometers. This is in part due to the advent of charge-coupled devices (CCD's) and other forms of array detector, which have different requirements to single-channel detectors such as photomultipliers; many details of the application of CCD's in spectroscopy are described by Sweedler, Ratzlaff and Denton (1994). Three forms of dispersive spectrometer are shown in figure 17. In the first case, with a single spectrometer, the light from the sample is focussed to an image in the plane of the entrance slit. As the name implies, a narrow slit, usually vertical, is placed here, and the light is directed so as to pass through this slit. The slit lies in the focal plane of the next component, which is a concave mirror; this mirror directs the light (now a parallel beam) on to the diffraction grating. Here, the different wavelengths of light are dispersed into parallel beams travelling in slightly different directions (the path of one such beam is drawn). This beam is directed towards a second concave mirror where it is reflected to form a monochromatic image of the entrance slit in the plane of the exit slit. Light of nearby wavelengths forms neighbouring line images in this plane, so that the width of the exit slit determines what range of wavelengths emerges from the instrument. In this way, the slit width determines the resolution of the spectrometer.

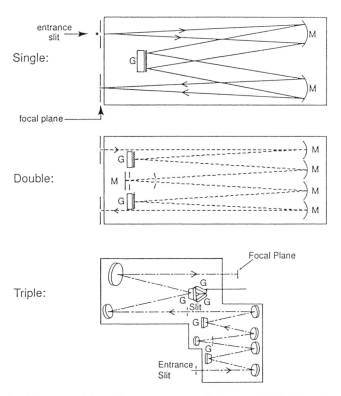

Figure 17. Single, double and triple grating spectrometers. Gratings are labelled G and mirrors M. With pemission, Sweedler et al. (1994).

Essentially, a single spectrometer can have a high enough resolution for Raman spectroscopy but generally it will not have good enough stray light rejection. The use of a double spectrometer improves the stray light rejection at the expense of efficiency (more mirrors and gratings lead to increased losses). In the double spectrometer, the two halves of the spectrometer are isolated from one another, the only optical path being through the narrow central slit. The likelihood of unwanted light being scattered by dust or imperfections through this instrument is much less than in a single spectrometer, so that the stray light rejection is good. However, the instrument is still only able to transmit a narrow wavelength band and is therefore only suitable for use with a single detector.

Nowadays, array detectors are available which measure many wavelengths simultaneously (arrays consist of typically around 1000 elements); the array detector is placed where the exit slit would normally be and is thus exposed to a series of images of the entrance slit, with images of different wavelengths falling on the array at different positions.

The triple instrument is specifically designed for array detectors. The light is passed through a wide central slit between the first and second stages and the gratings are set so that the *entire spectral range of interest* passes through this slit, but most of the Rayleigh-scattered light is blocked. The first and second stages are arranged so the dispersions of the two gratings are opposed; thus, the light entering the final stage is not dispersed. The final stage disperses this light once more, with as high a resolution as is required, and now the exit slit can be replaced by an array detector located in the focal plane. The characteristics of typical instruments of these types are summarised in the following table (adapted from Sweedler, 1994). Here, the symbol f# represents the ratio of focal length to dispersing element diameter, and thus indicates the angular size of the cone of light that the spectrometer can accept; generally, smaller values of this number are more advantageous.

Table. Comparison of the performance of different types of spectrometer employed in Raman spectroscopy.

Instrument type	f#	transmission	dispersion (cm^{-1} per mm)	stray light rejection
Single: 0.64 m, 1800 g/mm (Instruments SA 640)	5.2	0.32	26	10^{-5}
Double: 0.85 m, 1800 g/mm (Spex 1403)	7.8	0.10	12.5	10^{-12}
Triple: 1800 g/mm last stage (Spex Triplemate)	9	0.11	35	10^{-14}
Echelle:	less than for single	less than for single	resolution 1 cm^{-1}	-
Filter-based: imaging mode uses filter only; e.g., Renishaw Transducer Systems.	-	> 0.25 with single grating	resolution about 1 cm^{-1} with grating or 20 cm^{-1} in imaging mode.	-

Two other recent developments are worth mentioning; these are the use of echelle gratings and of holographic "notch" filters. Echelle gratings (figure 18) are a special type of diffraction grating (used at very high diffraction orders) which have high dispersions but which suffer from the problem that the different diffraction orders overlap significantly. This problem is overcome if an array detector is used with a second, low-resolution grating at right-angles to the echelle grating, so that the different orders are displaced from one another. In figure 18, the echelle grating disperses light vertically and the cross-disperser

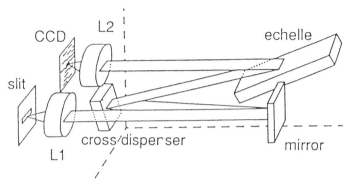

Figure 18. Optical arrangement of an echelle-based spectrometer with a CCD detector; L1 and L2 are lenses. From Sweedler et al. (1994) with permission.

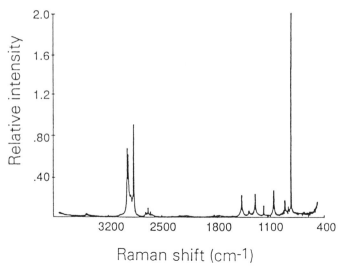

Figure 19. Raman spectrum of cyclohexane obtained with an echelle spectrograph and CCD detector, with permission from Sweedler et al. (1994).

disperses the different orders horizontally. With this system, the *entire* frequency range of vibrational modes can be covered at high resolution in one acquisition; figure 19 illustrates this with a spectrum of cyclohexane acquired in just one second (Sweedler et al., 1994).

Raman Imaging

One exciting recent development in Raman spectroscopy is the use of CCD's to form images of samples using the light emitted in a specific Raman band, so that regions composed of different materials can be distinguished. The key to this technique is the use of holographic band-pass filters which transmit the required wavelength band whilst providing excellent stray light rejection at other wavelengths. These filters are manufactured to operate at specific wavelengths and have very narrow band passes of down

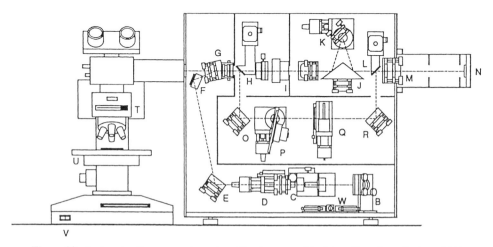

Figure 20. An imaging Raman spectrometer with microscope (Renishaw Transducer Systems Ltd.)

to 100 cm^{-1} (depending on the type of filter); their advantage over diffraction gratings is that two-dimensional spatial information is preserved.

Figure 20 shows an instrument (manufactured by Renishaw Transducer Systems) which uses a combination of holographic filters (labelled G and P) and a Fabry-Perot etalon (labelled Q) to achieve a spectral resolution of about 20 cm^{-1} in a true two-dimensional imaging mode (the CCD detector is labelled N on figure 20 and the relevant beam path is G-H-O-P-Q-R-L-M-N). The instrument can also be used with a single diffraction grating to provide conventional spectra at higher resolution (the grating is at K). The instrument is combined with a microscope, as discussed below. One potential disadvantage of this instrument is the difficulty of using tunable lasers as excitation sources (since a continuous range of filters is then required).

Raman Microscopy

As figure 20 indicates, Raman spectroscopy can be combined with optical microscopy to investigate microscopic samples present as, for example, inclusions within other materials. This technique has found many industrial applications, for instance in the diagnosis of problems in the production of plastics, where catalyst particles embedded in the polymer can be identified.

A microscope may be used in conjunction with a Raman spectrometer in two basic ways. Firstly, one may select a region of a sample by inspecting its white-light image; the Raman spectrum of that region may then be recorded. Alternatively, one may (by using a spectrometer such as that in figure 20) record a two-dimensional "map" of the intensity of some Raman band of the material as a function of position.

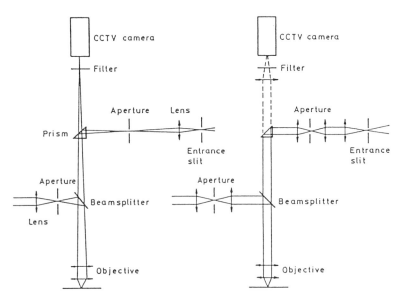

Figure 21. Optical layout of two types of Raman microscope, Gardiner and Graves (1989). In each case, the laser beam enters horizontally from the left and the Raman-scattered light exits horizontally to the right towards the spectrometer.

Figure 21 shows two possible optical arrangements of Raman microscopes; it is not possible here to give details of optical microscopy, but Gardiner and Graves (1989) give a very useful introductory review of this subject. In both (a) and (b) of figure 21, the laser beam is introduced into the microscope via a side-arm with a spatial filter to improve the beam profile; a beam splitter then reflects some of the laser beam down to the sample through the objective lens. The scattered light is collected through the same objective lens and some of it passes through the beam splitter. A movable prism or mirror can be inserted into the microscope column to reflect the beam out to a spectrometer; the prism can also be withdrawn so that the light passes up to a video camera in order to allow identification of the desired region of the sample. Spatial resolutions of around 1 µm can be achieved, depending on the particular optical system used. Note that by use of a confocal system with appropriate spatial filtering, it is possible to obtain a very restricted depth of field, so that spectra may even be obtained from different depths in transparent samples.

Optical Fibres

A common difficulty in industrial applications of Raman spectroscopy is that it may not be convenient to examine samples in the laboratory, either because frequent sampling is required or because the material of interest must be examined under conditions of extreme temperature, pressure, or chemical environment. Optical fibres provide a means of delivering the laser light to the sample and of collecting the scattered light. A system for such measurements is shown in figure 22; a single fibre is used to transmit the excitation light, whereas the scattered light is collected by several fibres formed into a "bundle". A lens is used to image the light emitted from the bundle into a conventional Raman spectrometer; often, the bundle is shaped into a line of fibres which match the size of the spectrometer slit. A disadvantage is that the glass of the fibre carrying the laser light will

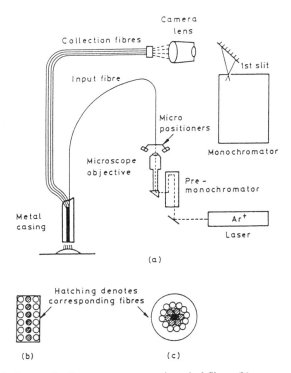

Figure 22. (a) A typical system for Raman spectroscopy via optical fibres; (b) arrangement of the collection fibres in front of the camera lens: (c) view of the end of the probe showing the input fibre (the central, solid black dot), Gardiner and Graves (1989).

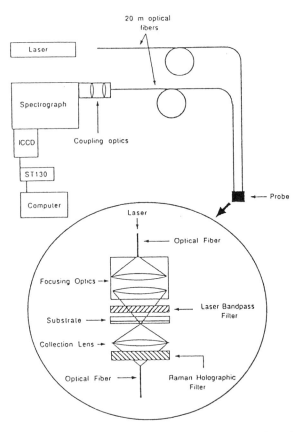

Figure 23. Fibre optic Raman system designed to minimise Raman scattering from the fibres, from Alarie et al. (1992).

itself give rise to some Raman scattering. Background subtraction procedures can solve this, but another approach is shown in figure 23, where a laser bandpass filter before the sample blocks the Raman-scattered light generated by the input fibre, whilst the holographic filter placed after the sample prevents stray excitation light from entering the collection fibre.

Surface-enhanced Raman Scattering

Finally, surface-enhanced Raman scattering (SERS) deserves mention. The effect was a subject of much research activity when it was first discovered; the field has been reviewed recently (Cardona and Güntherodt, 1983) and there are several textbooks on the subject. Essentially, a large enhancement of Raman-scattering cross-sections is sometimes observed for samples supported on roughened or corrugated metallic surfaces (frequently silver is used). One origin of this effect is the increased electric field of the excitation light near the surface, which results when the incident photons interact with charge density oscillations (surface plasmons) of the metal layer; the role of the roughening of the surface is then to relax the requirement of wavevector conservation and thus allow the transfer of energy from the incident light to the surface plasmons. However, other mechanisms are also believed to be of importance and a full discussion cannot be given here.

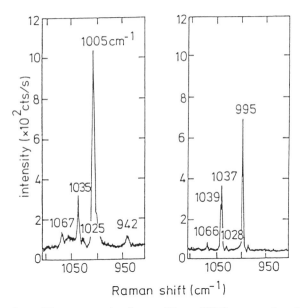

Figure 24. Surface-enhanced Raman scattering from pyridine at 120 K on a roughened silver surface (left) and a smooth silver surface (right). The absolute signal intensities are comparable but the concentration of pyridine in the left-hand case was 10,000 times smaller than in the right-hand case, showing the enhancement on the rough surface.

The effect is now often used as a technique for increasing the sensitivity of Raman spectroscopy, especially where only small quantities of samples are available. An illustration of the effect is given in figure 24, where a comparison is made between the spectra of approximately a monolayer of pyridine on an "activated" silver surface and a thick layer of pyridine on an inactive surface. Similar intensities are obtained from very different quantities of material. Figure 24 also demonstrates that the interaction with the surface can also modify the form of the spectrum. The nature of the spectrum may also be influenced by the electrical potential of the silver surface; the interaction between adsorbate and metal surface is a complex one so that this is not necessarily a simple technique to apply. Recently, SERS has been combined with microscopy in the use of extremely fine silver wires as probes which can selectively enhance signals from the very small region of the sample surrounding the tip of the wire; this has found application in biological systems (Alarie et al., 1992).

SUMMARY AND CONCLUSION

This chapter has dealt with the classical theory of Raman scattering and has given an indication of how the selection rules for Raman scattering arise. Some brief comments on the quantum mechanical theory have also been given. A series of case studies from the work of the author's group has been used to illustrate the range of topics that can be studied and the nature of the information that can be gained. Finally, some recent developments in experimental methods have been discussed. In a short review such as this, it is impossible to give a complete picture of the wide range of Raman spectroscopic techniques, or the enormous variety of research problems in which Raman spectroscopy can play a role. References have been given to more thorough treatments of all the above topics.

However, the field of Raman spectroscopy continues to develop rapidly and journals such as *Applied Spectroscopy*, the *Journal of Raman Spectroscopy* and *Physical Review B* regularly contain articles on new applications of Raman spectroscopy; there are also international conferences dedicated to this field whose proceedings are also informative. For an up-to-date introduction to Raman spectroscopy as applied to problems in chemistry, the book by Gardiner and Graves (1989) is very helpful, whilst the series of volumes edited by Cardona and Güntherodt (1983) is an invaluable reference work for recent applications of Raman spectroscopy in condensed matter physics.

ACKNOWLEDGEMENTS

The work of the Optical Spectroscopy group at UEA has been supported by the Engineering and Physical Sciences Research Council, the British Council, the Nuffield Foundation, the Royal Society, the European Union (Human Capital and Mobility programme) and the Deutscher Akademischer Austauschdienst.

REFERENCES

Alarie, J. P., Stokes, D. L., Sutherland, W. S., Edwards, A. C., and Vo-Dinh, T., 1992, *Appl. Spectr.* 46:1608.

Anderson, A., (ed.), 1971, "The Raman Effect," Dekker, New York.

Boardman, A. D., O'Connor, D. E., and Young, P. A., 1973, "Symmetry and its Applications in Science," McGraw-Hill, London.

Bower, D. I., and Maddams, W. F., 1989, "The vibrational spectroscopy of polymers," Cambridge Solid State Science Series, Cambridge Univ. Press.

Boyce, P. J., Davies, J. J., Wolverson, D., Ohkawa, K., and Mitsuyu, T., 1994, *Appl. Phys. Letts.* 65:1.

Cardona, M., and Güntherodt, G., (eds.), 1983, "Light Scattering in Solids I - VI," Topics in Applied Physics vols. 8, 50, 51, 54, 66 and 68, Springer-Verlag.

Demtröder, W., 1988, "Laser Spectroscopy," Springer Series in Chemical Physics vol. 5, Springer-Verlag, Berlin.

Gardiner, D. J., and Graves, P. R., (eds.), 1989, "Practical Raman Spectroscopy," Springer-Verlag, Berlin.

Halsall, M. P., Wolverson, D., Davies, J. J., Lunn, B., and Ashenford, D. E., 1992, *Appl. Phys. Letts.* 60:1.

Halsall, M. P., Railson, S. V., Wolverson, D., Davies, J. J., Lunn, B., and Ashenford, D. E., 1994, *Phys. Rev. B.*, 50:11755.

Harju, M. E. E., Valkonen, J., Jayasooriya, U. A., and Wolverson, D., J., 1992, *Chem. Soc. Faraday Trans.* 88:2565.

Hecht, J., 1986, "The Laser Guidebook," McGraw-Hill, New York.

Kuzmany, H., 1989, "Festkörperspektroskopie: Eine Einführung," Springer-Verlag, Berlin.

Loudon, R., 1978, "The quantum theory of light," Oxford University Press, Oxford.

Pelletier, M. J., 1990, *Appl. Spectr.* 44:1699.

Phillips, R. T., Wolverson, D., Burdis, M. S., and Fang, Y., 1989, *Phys. Rev. Letts.* 63:2574.

Porto, S. P. S., Giordmain, J. A., and Damen, T. C., 1966, *Phys. Rev.* 147:608.

Railson, S. V., Wolverson, D., Davies, J. J., Ashenford, D. E., and Lunn, B., 1993, *Superlattices and Microstructures* 13:487.

Sweedler, J. V., Ratzlaff, K. L., and Denton, M. B., 1994, "Charge-transfer devices in spectroscopy," VCH Publishers, UK.

INDUSTRIAL APPLICATIONS OF RAMAN SPECTROSCOPY

Neil J Everall

ICI Wilton Research Centre,
PO Box 90
Wilton
Middlesbrough,
Cleveland
TS90 8JE

INTRODUCTION

Raman spectroscopy is a multi-faceted technique which, in its various forms, utilises all of the properties of lasers that make them valuable as excitation sources for spectroscopy. For example, the monochromaticity, spatial coherence, polarisation, tunability, brightness and ability to run in pulsed mode are all vital properties that are exploited in various forms of Raman spectroscopy. Furthermore, the ability to collect data in a non-contacting manner over long distances, using fibre-optic probes, is becoming important for remote sampling and process monitoring. Raman spectroscopy probably represents the single largest application of laser spectroscopy in industrial analysis, a position which is expanding with the improved performance available from modern spectrometers, filters and detectors. This is reinforced by the unique capability to obtain Raman data from almost any sample irrespective of its physical form, in contrast to a great many spectroscopies which place constraints on sample presentation or preparation.

In this paper an attempt is made to illustrate the wide range of applications in the chemical and materials industries, drawing primarily on examples from our own laboratory. These include the identification of microsamples and contaminants, measurement of solid state morphology, quantification of polymer composition, measurement of reaction kinetics, the study of polymer degradation, and the use of Raman spectroscopy for process analysis. In an article of this length it is not possible to treat these subjects in depth, but a more detailed discussion of some specialised analytical applications can be found elsewhere (Grasselli and Bulkin, 1991). The theoretical and experimental aspects of Raman spectroscopy are covered by Wolverson elsewhere in this volume, and also a number of excellent sources cover these topics in detail (Long, 1977; Gardiner and Graves, 1989). Only those aspects that impinge directly on Raman as an industrial analytical technique are discussed in detail here.

An Introduction to Laser Spectroscopy
Edited by David L. Andrews and Andrey A. Demidov, Plenum Press, New York, 1995

INSTRUMENTATION FOR RAMAN ANALYSIS

The nature of the (linear) Raman effect places severe demands on the instrumentation used for its detection. It is an extremely weak effect; the oft-quoted example is that if ca 10^8 laser photons are incident on a sample, only *one* will reach the detector with a Raman shift, the rest will be scattered elastically and simply contribute noise. (For stimulated Raman scattering, a much higher fraction (up to ca 50%) of photons can be Raman converted; however, while these techniques do have industrial applications, we concentrate solely on spontaneous Raman scattering in this chapter). The first requirement is therefore to remove the elastically scattered (Rayleigh) photons from the signal of interest, and to analyse the frequency-intensity spectrum of the Raman scattered photons. This can be achieved using either interferometric or dispersive technology (Gerrard and Bowley, 1989). In fact, despite the weakness of the effect, it is now possible using spectrograph - Charge-Coupled Device (CCD) detector combinations to achieve unenhanced detection of material monolayers on dielectric substrates (Yarwood, 1994).

The second problem is that as well as scattering, laser absorption and re-emission (fluorescence) is prone to occur, particularly with the contaminated or recycled samples common in industrial environments. This fluorescence is often many orders of magnitude more intense than the Raman scattering, which is then masked by the background noise. A wide range of methods, ranging from temporal resolution to polarisation modulation, have been proposed for reducing fluorescence backgrounds; these have been comprehensively reviewed elsewhere (Everall, 1986). However, by far the most successful is to use a red or near infrared (NIR) laser beam in which the photons have insufficient energy to excite an electronic transition, thereby avoiding fluorescence. This is commonly achieved with a Nd:YAG laser emitting at 1.064 µm. Unfortunately, detectors for the NIR (~1µm) tend to have high background noise levels compared with the shot noise on the Raman photon flux, therefore they are normally coupled to an interferometer to allow improvements in signal to noise ratio (S/N) from the multiplex (Fellgett) and throughput (Jacquinot) advantages. This has given rise to the field of Fourier-transform (FT) Raman spectroscopy, which has revolutionised Raman spectroscopy and greatly broadened the user base in both industry and academia (Hendra et al., 1991).

It should be realised that the advantage of FT-Raman spectroscopy lies in reducing fluorescence through the use of NIR excitation; it is not a *sensitive* technique, because (1) the use of an infrared laser reduces the Raman signal, the scattering efficiency being approximately dependent on the fourth power of the laser frequency (Long, 1977), and (2) the detector noise is large compared to visible photodetectors. The effect of the relative magnitudes of detector and photon noise on the performance of FT and dispersive systems has been discussed elsewhere (Everall and Howard, 1989). As an alternative, the use of single spectrograph dispersive stages and holographic filters to reject the Rayleigh line, (Tedesco et al., 1993), with CCD detection, is a much more sensitive approach to Raman spectroscopy (Williams et al., 1990) although the maximum useful laser wavelength with these configurations is currently ca 800 nm. It remains to be seen if this wavelength is sufficiently long to eliminate fluorescence from the majority of "difficult" samples. At present, the industrial analyst ideally requires two systems, a 1.064 µm -excited system (probably FT Raman) to study highly fluorescent samples, and a dispersive CCD system (visible - ca 800 nm excitation) for cases where the utmost sensitivity is required. For example, practical Raman microscopy at spatial resolutions of ca 1 µm can currently only be achieved realistically using dispersive technology. The discrepancy in sensitivity is dramatically demonstrated in the Process Analysis section of this chapter. Figure 1 illustrates schematically the two approaches to Raman spectroscopy discussed above.

In the work described in this chapter, all (laboratory) dispersive spectra were recorded using a Dilor XY Raman spectrometer using 488 nm or 514.5 nm (argon ion laser) excitation. FT

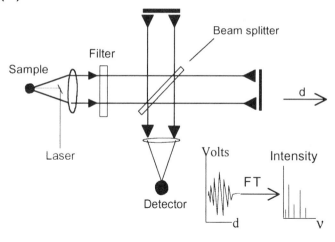

(a) FOURIER TRANSFORM RAMAN SCHEMATIC

Beam splitter

Filter

Sample

Laser

Volts

FT

Intensity

Detector

d

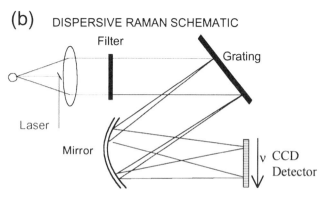

(b) DISPERSIVE RAMAN SCHEMATIC

Filter

Grating

Laser

Mirror

ν CCD
Detector

Figure 1. A schematic comparison of interferometric and dispersive Raman spectrometers. In each case the performance of the pre-filter in removing the Rayleigh scattered photons is vital. This is commonly achieved using holographic filters. Prior to this advance, a double monochromator would normally have been required for the dispersive system as a pre-filter. The mirror displacement for the FT instrument is denoted "d".

Raman spectra were obtained using a Perkin Elmer 1760 instrument (1064 nm Nd:YAG laser) and Process Analysis data were recorded using a Kaiser Optical Instruments Holoprobe 532 spectrometer using 532 nm excitation (frequency doubled Nd:YAG) and a fibre optic probe.

BENEFITS (AND DISADVANTAGES) OF RAMAN AS AN INDUSTRIAL ANALYTICAL TECHNIQUE

The major disadvantages of Raman spectroscopy have already been discussed above, namely the weakness of the effect and the influence of fluorescence. In additon, samples can degrade in the laser beam, or change morphology, or simply heat up and incandesce. This is a particular problem for FT Raman spectroscopy where incandescence occurs in the NIR, thereby masking the Raman signal. Furthermore, the small laser spot sizes on the sample (1 mm - 1 μm) can result in non-representative spectra of inhomogeneous samples. However, there are also some significant *advantages*.

(1) Since Raman scattering occurs through a modulation of the sample polarisability by a molecular vibration, as opposed to a change in dipole moment, it can be sensitive to different vibrations and species compared with those giving strong infrared (IR) absorption bands. For example, species such as C=C, C≡C and S-S give strong Raman bands, while O-H, C=O, N-H are strong in the IR. Some species (eg aromatics, nitriles etc.) give strong bands in each effect. In the extreme case of a molecule or crystal with inversion symmetry, no fundamental bands can be simultaneously Raman and IR active. Thus for complete characterisation of a material, both techniques would ideally be applied.

(2) Both glass and water are weak absorbers in the visible, and also weak Raman scatterers. Therefore Raman can be readily applied to study aqueous media, or samples stored in glass vessels, or glass fibre optics can be used for remote sampling. This contrasts sharply with IR absorption, where special optical materials are required, and aqueous samples present some difficulty. It should be noted that NIR-excited Raman spectroscopy can suffer from attenuation of Raman bands by the water absorption in the NIR.

(3) Since Raman is a scattering technique, samples can be examined as received, without preparation as a thin film, ground disk or mull. This is advantageous where preparation might change the sample prior to analysis. Samples can be solids (single crystals, amorphous, or polycrystalline powders), liquids or gases.

(4) Utilising a Raman microscope, samples can be analysed with micrometer resolution (roughly an order of magnitude improvement over that obtainable in the mid-IR).

(5) Resonance Raman scattering can detect low levels of chromophores.

(6) Inorganic species often give sharp Raman bands rather than broad features that can mask large regions of the IR spectrum. In addition, Raman provides easy access to the low frequency vibrations (≤ 400 cm^{-1}) important for characterising lattice vibrations and sample morphology.

We now turn to some examples of how these attributes of Raman spectroscopy are applied in industrial analysis.

IDENTIFICATION

The use of vibrational spectroscopy for identification of unknown materials relies on the fact that the number, frequency and intensity of vibrational bands depend upon the chemical structure and geometrical configuration of a given molecule or lattice, with the result that the spectrum acts as a fingerprint for the species concerned (and often its environment and morphology). While certain bands can be assigned to well defined groups (eg C=O or C-H stretching), others arise as a complex combination of bond distortions which are unique for a given structure. While this makes prediction of the vibrational spectrum of a complex molecule from first principles very difficult, it allows unknown species to be reliably identified by comparison with reference spectra.

This is illustrated by Fig. 2, which shows the Raman spectrum of a heterofilament polymer fibre recorded using a Raman microscope. By focusing the laser beam on to the outer sheath of the fibre (ca 3 μm thick), a spectrum characteristic of polyethylene was obtained. Focusing into the core of the fibre gave a spectrum predominantly indicative of polypropylene. In this case there was some contamination of the polyethylene spectrum by that of polypropylene, indicating that the spatial resolution was actually poorer than 3 μm in this case. Because the Raman microprobe used for this work was not configured for true confocal operation (Meier and Kip, 1994), the spatial resolution was degraded by spreading of the focused laser beam beyond the diffraction limit (Gardiner et al., 1986). However, it illustrates the use of the method for identifying unknowns. Further examples have been given by Louden (1989), and a detailed treatment of Raman microscopy was published by Rosasco (1980).

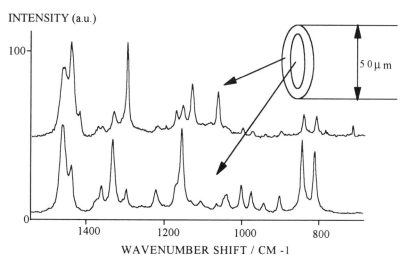

Figure 2. Identification of core and sheath components of a heterofilament polymer fibre using a Raman microprobe. The polymer was ca 50 μm diameter with a 3 μm sheath.

Raman microprobes can also be employed to obtain spectral images or maps, either by forming a two-dimensional image at a specific Raman frequency on the CCD chip (Garton et al., 1993) or by generating a laser line focus on the sample, and imaging the spectrum obtained from this line on to the CCD to obtain spectra simultaneously from points along this line (Bowden et al., 1990). Raman imaging is finding a niche as a quality control tool for the semiconductor industry, for example in checking the quality of hard disks, or the quality of diamond-like carbon coatings. Another example is mapping distributions of components in polymer blends. Point-by-point mapping using a motorised translation stage is an obvious alternative method of obtaining a spectral map, although this can be time-consuming .

QUANTITATION

Raman spectroscopy can be used for quantitation since, to a first approximation, the intensity of a Raman band is proportional to the concentration of the vibrating species. In fact, the absolute intensity is also dependent on factors such as the laser power, sample refractive index, and changes in scattering cross-section due to intermolecular interactions. Furthermore, because Raman spectroscopy is a "single beam" technique, the optical sampling geometry and spectrometer efficiency will also influence the detected intensity, to the extent that the measured band intensity usually has little physical meaning. Although absolute Raman intensities have been recorded, this is a difficult process which is exceptionally time-consuming. For this reason it is usual to normalise band intensities relative to an internal standard (eg a solvent peak) that is recorded simultaneously with that of the analyte.

Figure 3 illustrates this approach, which shows the analysis of a copolymer of styrene and butylacrylate used as a rubber. The FT-Raman spectrum clearly shows discrete features due to the acrylate (C=O stretch) and styrene groups (aromatic ring stretch). By measuring the spectra of samples of known composition, it is possible to construct a calibration relating the ratio of the intensities of the styrene and carbonyl peaks to the styrene / acrylate mass ratio (see inset). This allows the composition of an unknown sample to be readily determined from its spectrum. In addition, the presence of a small peak which can be assigned to allyl methacrylate is seen in the spectrum, which demonstrates that this trace co-monomer has not been fully consumed in the expected cross-linking reaction. The main benefit of the analysis is that it can readily be carried out on cross-linked rubbers which are insoluble and hence intractable to analysis by solution state nuclear magnetic resonance (NMR) or traditional IR methods (which would also be insensitive to residual allyl species).

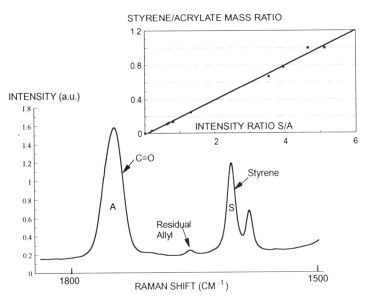

Figure 3. Analysis of co-monomer composition of a cross-linked styrene - butyl acrylate rubber. The ratio of the peak intensities (S/A) correlates linearly with the styrene/acrylate concentration ratio.

Another example of quantitation using Raman spectroscopy is given in Fig. 4. This shows monitoring of the thermal polymerisation of a cyclic phosphazine to form a linear product. In this case both the starting material and the product were air-sensitive, so the reaction was carried out under an inert atmosphere in a sealed glass vessel. Raman is an ideal technique for monitoring this reaction since it is possible to focus the laser beam through the glass wall and collect the backscatter in a non-invasive manner. Conversion of the monomer is monitored by observing the growth of a band at ca 450 cm^{-1}, assigned to P-Cl stretch in the linear polymer, relative to the P-Cl stretch in the cyclic monomer (ca 360 cm^{-1}). Once calibrated using samples of known composition (obtained by terminating the reaction, functionalising the sample and performing NMR), the technique can be used to follow the progress of the reaction and indicate the optimum time to terminate heating. It is difficult to envisage another technique which could yield the same information non-invasively.

Other examples of the use of Raman spectroscopy to monitor polymerisation have been reported, for example in the polymerisation of styrene and acrylics (Gulari et al., 1982). This field should expand with the growth of process Raman spectroscopy as a viable technique.

The above examples illustrate the principles of quantitative analysis using Raman spectroscopy. These are discussed in more detail by Vickers and Mann (1991). In addition, only the use of a single band parameter (e.g. a height or integrated intensity) correlated against a calibration value such as a concentration was considered. However, multivariate (including Factor Analysis) calibrations, in which changes at many points in the spectrum are simultaneously correlated with calibration values, are becoming more widely used in Raman spectroscopy. These

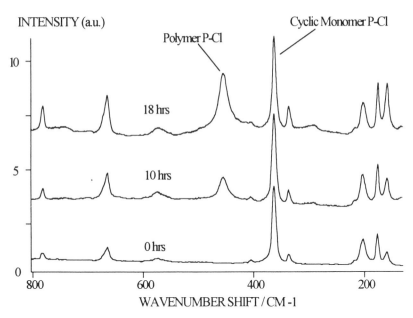

Figure 4. Monitoring phosphazine polymerisation non-intrusively as a function of time.

are typically more robust and precise than univariate calibrations. Examples have been given by Irish and Ozeki (1991) for analysing inorganic equilibria, and Everall et al. (1994) for quantifying polymer morphology.

MORPHOLOGY

General

For the purposes of this work we consider morphology to include the characterisation of crystal structure, polymorphism, phase composition, degree of crystallinity of semi- crystalline materials (eg polymers), molecular conformation, and crystallite size. As well as analysing bulk materials it is possible, by utilising the Raman microprobe, to characterise small particles and to map morphological properties on the microscopic scale (Chalmers et al., 1990). We neglect here the important field of measuring molecular orientation functions, an area in which analysis of the intensity and polarisation characteristics of Raman bands can yield much useful information (Bower, 1972).

Sample morphology influences the vibrational spectrum in several ways. Changes in molecular conformation or crystal form will change the symmetry of the material (either molecular or crystal point group) and this can influence the number and Raman activity of the normal modes of vibration. It is possible to predict the influence of such changes for molecules or crystals using group theory (Cotton, 1971; Fateley et al., 1972). Secondly, changes in molecular conformer population (for example, in a polymer) can cause changes in intensities of bands due to specific conformers, or bandshape changes. This has been illustrated, for example, for poly(ethylene terephthalate) (PET) undergoing crystallisation (Melveger, 1972; Adar and Noether, 1985). Thirdly, reduction of crystal size can invalidate the selection rules derived on the assumption of an infinite crystal lattice, thereby introducing new bands or band broadening, or bands due specifically to surface modes. This has been observed, for example, for graphitic carbon (Tuinstra and Koenig, 1970).

Analysis of Lattice Modes

One of the main advantages of Raman spectroscopy for analysing morphology is that the sample can be examined in its native form without preparation. This eliminates the possibility of inadvertently modifying the sample prior to analysis. As an example, consider the analysis of terephthalic acid (TA). This material is an important precursor for PET, which in turn is a very important polyester used in manufacturing films and fibres. The acid exists as a powder which can contain two different crystal forms, denoted I and II, both of which display strong intermolecular hydrogen bonding. The flow properties of the powders depend strongly on the crystal form, and a method of quantifying the composition is sometimes needed in order to understand product properties.

Figure 5 illustrates the low frequency Raman spectra of (a) form I and (b) form II TA, obtained from the powder contained in a glass bottle without any preparation. Clearly, the

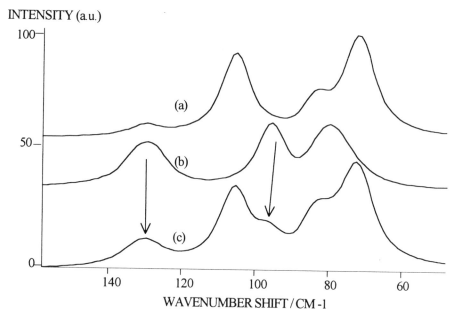

Figure 5. Use of lattice vibrations to distinguish crystal forms of terephthalic acid. Note the partial conversion of form I (a) into form II (b) under light hand pressure (c).

low-frequency, Raman-active vibrations of the crystal lattice can readily differentiate the two forms. In addition, the spectrum of a sample of form I TA which was lightly pressed as a pellet (using hand pressure) is also shown as spectrum (c). This spectrum is clearly a superposition of spectra (a) and (b), illustrating that form I can, at least partially, convert into form II under mild physical stimulation. In fact, simply vigorously shaking a bottle of the powder can cause the conversion to occur. This demonstrates the requirement for a method which does not require any sample preparation at all, in contrast to the need to press a disk for IR or powder X-ray analysis.

Another example of where this non-destructive aspect of Raman spectroscopy is advantageous is in the analysis of polymer crystallinity and molecular orientation, which can be carried out without the pressing, sectioning or grinding of samples. These procedures could easily disturb the morphology prior to analysis. Similarly, valuable samples such as large single crystals can be examined non-destructively.

Analysis of Crystallite Size: - Carbon

Elemental carbon is an important material, not least as a constituent of modern composites and hard coatings, so methods for its characterisation are important. Raman spectroscopy is a powerful technique in this area, perhaps surprisingly since one might expect such a dark material to burn up in the laser beam. Although this is certainly the case with NIR excitation, carbon is a strong Raman scatterer and use of a visible laser normally yields excellent Raman spectra that provide a great deal of morphological information. As such, this is a widely used tool for characterising carbon powders, fibres, composites and diamond-like films.

As an example, Fig. 6 shows the Raman spectra (514 nm excitation) of three samples of graphitic carbon of varying crystallite diameter. As the hexagonal lattice planes become smaller, a new band at ca 1350 cm^{-1} appears, and the band at 1580 cm^{-1} (present in all graphites) broadens and becomes weaker. Tuinstra and Koenig (1970) showed that the intensity ratio I(1350)/I(1580) varies linearly with the inverse of the lattice diameter L_a, as determined from X-ray line broadening, thus providing a measure of the crystallite size. The higher frequency band was assigned to the E_{2g} symmetry species, and the 1350 cm^{-1} band to an A_{1g} mode that is formally forbidden under the crystal symmetry of the graphite lattice. It was proposed that the reduction in transational symmetry on reducing crystal size allows the normal selection rules to be broken, and the A_{1g} mode becomes active. This is the usually accepted explanation for the band appearance and its intensity variations with crystal size. In addition, the width and position of the E_{2g} mode have been correlated with the distance between graphite layers, allowing characterisation of three-dimensional ordering in graphites (Katagiri, 1988).

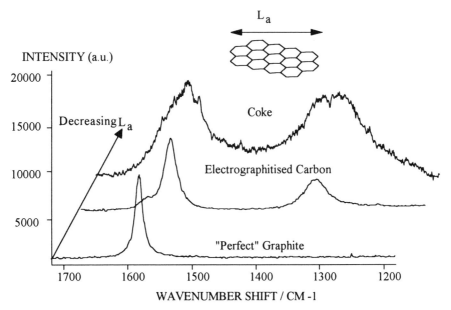

Figure 6. Variation in the Raman spectra of graphitic powders as a function of lattice perfection (i.e. crystallite diameter L_a). Note the growth of the new band with decreasing crystallite size.

The powder spectra in Fig. 6 arise from isotropic arrangements of crystallites in the laser beam, and as such would appear the same irrespective of the sampling geometry. This is not always the case; for example, Fig. 7 shows spectra obtained using a Raman microprobe from the polished end and the surface of a poly(acrylonitrile) (PAN) based carbon fibre of 7 μm diameter. The fact that the spectra are so different immediately indicates that the fibre structure is anisotropic; they can be interpreted primarily on the basis of a graphitic-like structure, but with the basal planes arranged with their normals perpendicular to the fibre long axis. This means that irradiation of the fibre surface predominantly samples basal planes and so yields the spectrum of a "good" graphite, whereas focusing the microprobe on the fibre end preferentially samples plane "edges", giving the appearance of smaller-crystallite material (Fitzer, 1988). The utility of the Raman microprobe for resolving the morphology of small structures is convincingly demonstrated by this example.

Some authors have attempted to correlate the E_{2g} band position with the degree of strain in carbon fibres, particularly when incorporated in a composite (for example, Robinson et al., 1987). This frequency shift arises from deformation of the lattice structure, perturbing the vibrational energy levels, and has been observed in a wide range of polymeric materials as well as carbon fibres. However, the shift in band position with strain is rather small, and it has been shown that with carbon fibres small fluctuations in laser power can induce variable sample heating and much larger bandshifts if care is not taken (Everall and Lumsdon, 1991). Furthermore, it was also shown by the above authors that the natural variation in E_{2g} mode frequency along individual unstrained carbon fibres could be sufficient to mask the expected variation due to strain in the composite. Thus while observation of a shift in mode frequency from a single point on a carbon fibre under controlled deformation conditions may be valid, the extension of these results to mapping strain in composite articles is fraught with problems that are usually ignored and might cast doubt on the conclusions drawn.

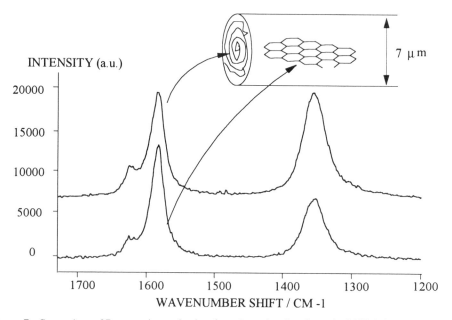

Figure 7. Comparison of Raman microprobe data from the end and surface of a PAN-derived carbon fibre. Note the clear structural anisotropy.

POLYMER DEGRADATION

Raman spectroscopy would not normally be considered a good tool for studying polymer degradation, since the latter often procedes via an oxidative route, and oxygenated species are not usually strong Raman scatterers. Furthermore, degradation often yields discoloured materials that are prone to fluorescence, making Raman measurements difficult. However, there is one classic example where Raman is a prime technique for analysing degradation, namely that of poly(vinyl

124

chloride), or PVC. This polymer degrades by loss of HCl to yield conjugated C=C bonds in the polymer backbone, $[-(C=C)_n]$, termed polyenes. The electronic absorption frequency strongly depends on the conjugation length n; once n exceeds ca 6 units, the polymer absorbs in the visible region of the spectrum, with an absorption maximum λ_{max} which depends on n. This is the phenomenon responsible for the discolouring of PVC window and door frames. The degradation can be controlled through the use of additives, but only low levels of polyene (ca 0.0001%) are needed to discolour the PVC. Therefore, methods for characterising the concentration and conjugation length of the polyene as a function of polymer formulation and degradation conditions are needed.

Raman spectroscopy provides just such a technique. When the excitation laser wavelength falls within an electronic absorption band of a sample, it is sometimes possible to observe "resonance Raman" scattering due to vibrations of the chromophore (Long, 1977). In this event, the change in polarizability during a vibration, and hence the scattering efficiency, can be enhanced by many orders of magnitude, allowing very low levels of chromophore to be analysed. Observation of this effect is not possible for all samples; for example, it is necessary that the sample does not degrade or fluoresce very strongly as it absorbs the laser light, but under suitable conditions it can be a very powerful technique. Fig. 8 illustrates the application to analysing PVC degradation at 80 °C as a function of time. As degradation procedes, the spectrum changes from that of "pristine" PVC (0 min) to include two new intense bands at ca 1500 and 1100 cm^{-1}, which completely dominate the spectrum after 20 minutes. These bands, normally termed v_2 and v_1 respectively, have been assigned to C=C and C-C stretching vibrations in the polyene, and it is known that the position of the v_2 band can be correlated with conjugation length "n" via Eq.1 (Baruya et al., 1983):

$$v_2(cm^{-1}) = 1461 + 151.24 \exp(-0.07808n) . \qquad (1)$$

Equation 1 demonstrates that the C=C frequency decreases as the conjugation length increases. It is interesting to note that although the polyene bands dominate the upper spectrum in Fig. 8, the level of polyene was probably only of the order of <0.0005%, illustrating the extreme sensitivity of resonance Raman scattering.

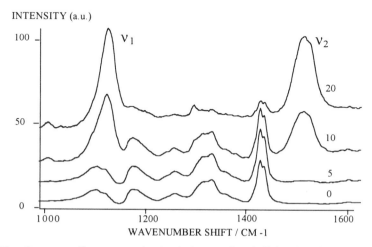

Figure 8. Use of resonance Raman scattering (excitation wavelength 514 nm) to detect polyene species in thermally degraded PVC. The position of the v_2 (equation 1) band indicates that the conjugated sequence length was ca 13 units. The degradation time (in minutes) is indicated on the spectra.

The intensities of the ν_1 and ν_2 bands in Fig. 8 can be used to compare the levels of polyene as a function of degradation time and temperature, and the ν_2 band frequency indicates a conjugation length of ca n=13 (equation 1). Gerrard and Maddams (1986) have discussed the use of such spectra to characterise in detail polyene generation and subsequent cross-linking. For example, by varying the laser excitation wavelength, different conjugation lengths will be brought into resonance, so a plot of ν_2 frequency versus excitation wavelength can give insight into the polydispersity of n. This is because longer sequences will have absorption maxima shifted to the red, so use of a red laser will preferentially detect longer conjugation lengths, with lower ν_2 frequencies. In the extreme, use of a 1.064 µm laser and an FT Raman spectrometer will detect exceptionally long sequences that are not observed under visible excitation (Williams and Gerrard, 1990). It is worth noting that resonance Raman spectroscopy has become one of the major tools for characterising conducting polymers in terms of conjugation (delocalisation) length and the nature of domains in which conduction electrons are localised (Furukawa et al., 1991).

PROCESS ANALYSIS

In principle, Raman spectroscopy is an attractive technique for on-line process monitoring owing to the availability of relatively cheap fibre optic materials for transporting visible and NIR radiation efficiently over long distances. However, until recently this has not been a viable technique owing to the complexity of Raman spectrometers. More recently, the advent of relatively low cost FT Raman and single-stage dispersive spectrometers has made plant-installed, robust Raman systems a possibility. For example, the use of FT Raman systems for monitoring distillation columns has been described by Garrison (1992).

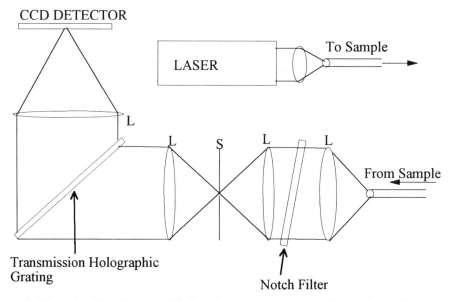

Figure 9. Schematic of the Holoprobe 532 dispersive Raman spectrometer (Kaiser Optical Systems, Ann Arbor, Michigan, USA). The heart of the unit is the transmission holographic grating (Fig. 10). L=lens, S=slit. Fibre optics (up to 100 m) are used to deliver the laser beam and Raman scatter to and from the sample.

We have recently been investigating the use of an integrated dispersive Raman system for on-line polymer analysis. This is a no-moving parts commercial spectrometer (Kaiser Optical

Systems Holoprobe 532) consisting of a diode-pumped, frequency doubled Nd:YAG laser (532 nm), a 100 m fibre optic probe, a holographic Rayleigh rejection filter, a transmission holographic grating and a CCD detector, all housed in a single unit which is shown schematically in Fig. 9.

The importance of the holographic filter for rejecting the Rayleigh scatter was mentioned earlier in this chapter; without it, a double monochromator filter stage would be required and the system size and cost would be drastically increased (and sensitivity decreased). However, the most important *novel* component is probably the transmission holographic grating. This actually contains two volume phase holograms (Tedesco *et al.*, 1993) which are configured such that one half of the Raman spectrum (ca -1000-1600 cm^{-1}) is imaged onto a one part of the CCD, while the remainder (1600-4000 cm^{-1}) is displayed on a lower area (Fig. 10). This means that the entire spectrum can be captured simultaneously at high spectral resolution, with no need to rotate the grating to "scan" the spectrum over the detector. This improves both the sensitivity of the system (i.e. there is a multiplex advantage to collecting all of the spectral elements simultaneously) and also its robustness and calibration stability.

The sensitivity of the system is illustrated in Fig. 11, which compares the spectra obtained from a polymer using the FT-Raman and dispersive systems. The dispersive spectrum was obtained through a 100 m fibre probe, using an order of magnitude shorter acquisition time and a 20-fold lower laser power, but still gave a S/N ca *twice* that obtained using the FT system and no

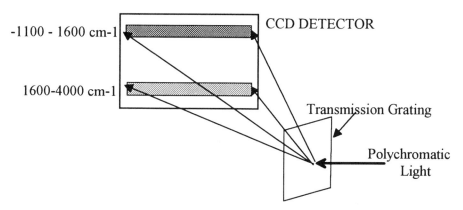

Figure 10. Schematic of multiplexed transmission holographic grating.

fibre probe. In fact, we have shown that in the absence of fluorescence, the dispersive system is generally ca three orders of magnitude more sensitive than the NIR-excited FT system using equivalent laser powers and acquisition times. The reason for this large difference is threefold; firstly, the scattering efficiency using a green laser is ca 20-fold higher than that using the NIR laser; secondly, the dispersive system is limited by the shot-noise on the Raman photon flux rather than that of the detector (Everall and Howard, 1989), and thirdly, the dispersive system is a high-throughput optical device which relays a large fraction of the collected photons to the detector. In fact, the sensitivity would be further improved by at least an order of magnitude if the system were directly coupled to a sample with a lens rather than the fibre optic probe, as would be possible for at-line, rather than on-line, *in situ* analyses.

As an example of the use of the dispersive system for an actual on-line measurement, Fig. 12 shows the spectrum of a semicrystalline polymer being produced under a range of uniaxial extensions. As the draw ratio varies, distinct changes in the spectrum are observed and which can be correlated with a *trans* conformation of the polymer backbone. This

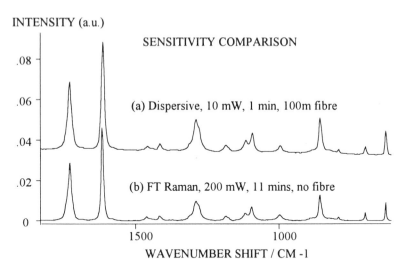

Figure 11. Comparison of the sensitivities of the FT-Raman and dispersive spectrometers. See text for details.

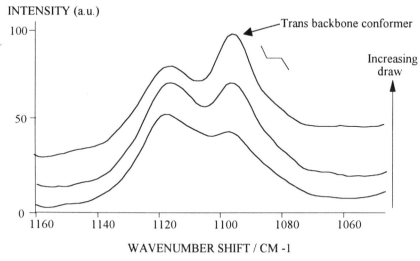

Figure 12. Use of process Raman spectroscopy to monitor polymer crystallinity during processing. Note growth of a band due to trans backbone conformation as polymer crystallises with increasing draw ratio.

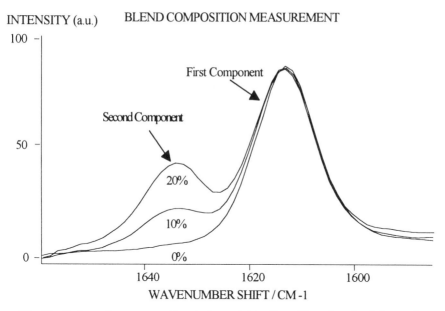

Figure 13. Monitoring polymer composition during melt-blending. The unique band due to the second (minor) component can be used to quantify its concentration.

corresponds to changes in the degree of crystallinity, an important property of the polymer, and allows monitoring and control of the polymer actually during production in order to optimise final properties. Figure 13 illustrates the monitoring composition of a polymer blend during production. In each case, spectra were recorded with ca 2 minutes elapsed time, through a 100 m fibre probe, illustrating the potential for real-time, remote *in-situ* analysis.

Process Raman spectroscopy is in its commercial infancy and is still somewhat inferior in its general applicability compared with, say, NIR absorption spectroscopy, probably the furthest advanced of the process spectroscopies. For the Raman process analyst there is often a conflict between the desire to use NIR laser-excitation to minimise fluorescence, and visible excitation to maximise sensitivity and Raman scatter, since many industrial processes will demand both fluorescence rejection and also sensitivity. The author believes that the way forward lies with spectrograph-array detection systems, (the 1.064 μm, FT Raman system is probably too insensitive for many process applications), but coupled to a laser probably operating up to ca 800 nm in order to minimise fluorescence.

Much effort is currently being directed into obtaining powerful, stable but compact red lasers for incorporation into dispersive process Raman systems; this is one of the keys to the success of the technique in this area. The same requirements exist for optimising highly sensitive, dispersive Raman microprobes for general use; a bright, frequency-stable red source is essential for each application.

CONCLUSIONS AND OTHER DEVELOPMENTS

It is hoped that the examples above give an impression of the wide range of applications which Raman spectroscopy finds in the industrial environment. The technical advances over the last 8 years have truly revolutionised the scope of the technique as an analytical tool; firstly, FT-Raman spectroscopy brought the technique into more widespread use as a result of the compactness, robustness and ease of use of the systems, as well as minimising the influence of fluorescence. More recently, the use of holographic optical elements, and CCD detectors superceding intensified diode arrays, has also resulted in simple and compact systems, both making Process Raman spectroscopy a more viable tool, and also producing extremely sensitive research-grade spectrometers and microprobes. The gains in sensitivity arise from the ability to work with a simple holographic filter, a single spectrograph and a high quantum efficiency, low noise CCD detector, thereby relaying a large fraction of Raman scattered photons to the detector while efficiently blocking the Rayleigh scatter.

The main technical areas in which improvements can still be made lie with the lasers and array detectors. Firstly, stable, high power red lasers (700-800 nm) are required to provide optimum excitation for CCD detectors. This is readily achieved in the laboratory, but compact lasers for use in portable Process Raman Analysers are not readily available. Secondly, low-noise array detectors with good response in the NIR (>1.1 μm) are required to allow the benefits of fluorescence rejection enjoyed by FT Raman systems to be realised in a dispersive, high sensitivity spectrometer. At present it is not clear "how red" we need to go to eliminate fluorescence from the majority of materials while still maintaining sensitivity; it may be that a CCD detector using excitation between 700-800 nm will be largely satisfactory. This is an area of intense experimental activity at present and the question may be resolved over the next few years. However, in our experience there are a significant number of materials which still either fluoresce or degrade even in a Nd:YAG laser beam, and moving to still longer wavelengths is not trivial. Raman spectroscopy will never, in the author's opinion, replace FTIR as the simple, laboratory based technique which will most often yield a vibrational spectrum from the majority of samples. Fortunately for the Raman spectroscopist, when a spectrum can be obtained, it can often enable measurements to be made which are impossible by other techniques!

Finally, it should be noted that this chapter has only given a flavour of the analytical applications of Raman spectroscopy; there are many important areas which it has not been possible to address, including Raman imaging / mapping, surface enhanced scattering, non-linear (stimulated) spectroscopies, catalysis (and measurements under extremes of temperature and pressure), measurement of molecular orientation, geochemical and biological applications, to name but a few.

ACKNOWLEDGEMENTS

The author is indebted to many of his colleagues at ICI for contributions towards the work described above, including John Chalmers, Janet Lumsdon, Steve Ragan, Bob Tooze and Duncan Mackerron. In addition, the help and advice of Perkin-Elmer (UK) and Dilor (France) is gratefully acknowledged. Harry Owen, Joe Slater and Mike Pelletier (Kaiser Optical Instruments, Ann Arbor, USA) are thanked for providing instrumental and technical help with our initial forays into Process Raman Analysis. ICI plc is thanked for granting permission to publish this work .

REFERENCES

Adar, F., and Noether, H., 1985, *Polymer* 26:1935.
Baruya, A, Gerrard, D.L., and Maddams, W.F., 1983, *Macromol.* 16:578.
Bowden, M., Gardiner, D.J., and Rice, G., 1990, *J. Raman. Spectrosc.* 21:37.
Bower, D.I., 1972, *J. Polymer Sci. B* 10:2135.
Cotton, F.A., 1971, "Chemical Applications of Group Theory," Wiley, New York.
Chalmers, J.M., Croot, L., Eaves, J.G., Everall, N.J., Gaskin, W.F., Lumsdon, J., and Moore, N., 1990, *Spectros. Int. J.* 8:13.
Everall, N.J., 1986, PhD Thesis, University of Durham, UK.
Everall, N.J., and Lumsdon, J., 1991, *J. Mater. Sci.* 26:5269.
Everall, N.J., Chalmers, J., Ferwerda, R., van der Maas, J., and Hendra, P., 1994, *J. Raman Spectrosc.* 25:43.

Everall, N.J., Tayler, P., Chalmers, J.M., Mackerron, D., Ferwerda, R., and van der Maas, J., 1994, *Polymer* 35:3184.

Everall, N.J., and Howard, J., 1989, *Appl. Spectrosc.* 43:778.

Fately, W.G., Dollish, F.R., McDevitt, N., and Bentley, F.F., 1972, "Infrared and Raman Selection Rules for Molecular and Lattice Vibrations: the Correlation Method," Wiley, New York.

Fitzer, E., and Rozploch, F., 1988, *High Temp.-High Press.* 20:449.

Furukawa, Y., Ohta, H., Sakamoto, A., and Tasumi, M., 1991, *Spectrochim. Acta* 47A:1367.

Gardiner, D.J., Bowden, M., and Graves, P.R., 1986, *Phil. Trans. R. Soc. Lond.* A320:295.

Garrison, A.A, Moore, C.F., Roberts, M.J., and Hall, P.D., 1992, *Process Control and Quality*, 1:281.

Garton, A, Batchelder, D.N., and Cheng, C., 1993, *Appl. Spectrosc.* 47:922.

Gerrard, D.L., and Bowley, H., 1989, Instrumentation for Raman Spectroscopy, *in*: "Practical Raman Spectroscopy," D.G. Gardiner and P. Graves, eds., Springer Verlag, Berlin-Heidelberg.

Gerrard, D.L., and Maddams, W.F., 1986, *Appl. Spectrosc. Rev.* 22:251.

Grasselli, J.G., and Bulkin, B.J., (eds.), 1991, "Analytical Raman Spectroscopy," Wiley-Interscience, New York.

Gulari, E., McKeigue, K., and Ng, K.Y.S., 1982, *Macromol.* 17:1822.

Hendra, P.J., Warnes, G., and Jones, C., 1991, "Fourier Transform Raman Spectroscopy," Ellis-Horwood, Chichester.

Irish, D.E., and Ozeki, T., 1991, Raman spectroscopy of inorganic species in solution, *in*: "Analytical Raman Spectroscopy," J.G. Grasselli and B.J. Bulkin, eds., Wiley-Interscience, New York.

Katagiri, G., Ishida, H., and Ishitani, A., 1988, *Carbon* 26:565.

Long, D.A., 1977, "Raman Spectroscopy," McGraw-Hill, New York.

Louden, J.D., 1989, Raman Microscopy, *in*: "Practical Raman Spectroscopy," D.G. Gardiner and P. Graves, eds., Springer Verlag, Berlin-Heidelberg.

Meier, R.J., and Kip, B.J., 1994, *Microbeam Analysis* 3:61.

Melveger, A.J., 1972, *J. Polym.Sci. A-2*, 10:317.

Robinson, I.M., Zakikhani, M., Day, R.J., Young, R.J., and Galiotis, C., 1987, *J. Mater. Sci. Lett.* 6:1212.

Rosasco, G.J., 1980 , *Adv. Infrared Raman Spectrosc.* 7:223.

Tedesco, J.M., Owen, H., Pallister, D.M., and Morris, M.D., 1993, *Anal. Chem.* 65:441A.

Tuinstra, F., and Koenig, J.L., 1970, *J. Chem. Phys.* 53:1126.

Vickers, T.J., and Mann, C.K., 1991, Quantitative analysis by Raman spectroscopy, *in*: "Analytical Raman Spectroscopy," J.G. Grasselli and B.J. Bulkin, eds., Wiley-Interscience, New York.

Williams, K.J.P., Dixon, N.M., and Mason, S.M., 1992, Proc. XIII Int. Conf. Raman Spectrosc., Wiley, Chichester, pp.1070-1071.

Williams, K.J.P., and Gerrard, D.L., 1990, *Eur. Polym. J.* 26:1355.

Yarwood, J., 1994, *personal communication*.

LASER REMOTE SENSING

Andrey A. Demidov

Moscow State University
Department of Physics
119899 Moscow, Russia

INTRODUCTION

This report is devoted to a brief survey of the major principles and ideas employed in LIDAR remote sensing. For more detailed information the book by Measures (1984) is recommended. Laser remote sensing is the sensing of spectroscopic parameters of various objects over distances ranging from meters to hundreds of kilometres. The device which allows such measurements to be conducted is called LIDAR. This is an abbreviation of *LIght Detection And Ranging*, after the well-known abbreviation RADAR. The basic principles of LIDAR and RADAR are the same, but RADAR uses a radio beam, while LIDAR employs light. The idea of applying laser sources for such remote detection was emphasised and employed in the 60's (Ligda, 1963; Hickman and Hogg, 1969; Inaba and Kobayasi, 1969), when lasers were discovered and first began to be applied.

Conventional light sources are unsuitable for a system like lidar, i.e. for remote sensing of the spectroscopic features of various species. The following specific properties of laser light are employed in numerous lidar systems: (a) high power and low divergence of the laser beam, which allows delivery of a large amount of light energy over long distances; (b) monochromaticity of

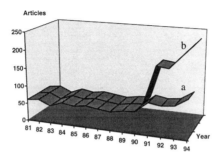

Figure 1. Number of articles about lidar sensing published yearly; graph (a) shows result of search of Science Citation Index (SCI) on articles' title, graph (b) shows the analogous kind of search which in addition covers abstract and keywords; the latter two have been included in SCI since 1991.

An Introduction to Laser Spectroscopy
Edited by David L. Andrews and Andrey A. Demidov, Plenum Press, New York, 1995

laser light, which allows selective excitation of the target; (c) laser pulses can be made very short (10^{-7}-10^{-12} sec), allowing - (*i*) avoidance of solar blindness, (*ii*) spatial selection of the target, and (*iii*) performance of kinetics measurements; (d) the coherence of laser light allows measurement of velocity via the Doppler effect; (e) multi-wavelength and multi-pulse laser beams open new areas of application, for example, in the remote sensing of biological productivity in natural waters.

This review will not attempt coverage of all known areas of lidar applications; it is impossible to do so in such a short report. According to the Science Citation Index more than 50 works devoted to lidar sensing are published each year (see Fig.1), and this number is constantly growing.

In general all lidar systems consist of the following principal components (see Fig.2): (a) the laser, which generates an incident light beam; (b) the optical system **O1**, which corrects the beam; (c) the optical system **O2**, which serves for collection and correction of the signal detected; and (d) the detector. In this figure **R** is the distance between the lidar and the target. The opti-

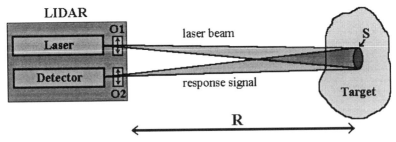

Figure 2. Lidar scheme.

cal system **O1** may contain, for example, an expansion telescope, which increases the laser beam diameter and consequently decreases its divergence. It may contain optical filters for selecting specific wavelengths of the incident light. The optical system **O2** may contain a simple lens or a sophisticated telescope as well as optical filters for collecting and filtering the information-bearing part of the response signal. The choice between lens or telescope depends on the ranging distance **R**. Thus, for distances of up to a hundred meters the simple lens with a diameter of about 10 cm would work with the same efficiency as a complicated telescope system. On the other hand, when the sounding distance is about 1 km or more a telescope with an effective diameter of about 25 cm or more would be optimal for the light collection.

When the sounding object is "hit" by a laser beam, it generates a response signal. It could be of various different natures: Rayleigh or Mie scattering, Raman scattering, fluorescence etc. The response signal contains information about the sounding target and its detection thus allows remote evaluation of the target parameters to be conducted. The simplified formula (the lidar equation) for the power of the detected signal P_{det} has the following form:

$$P_{det} = \frac{S}{\pi R^2} P_{las} \tau \varepsilon \gamma \qquad . \tag{1}$$

In this formula P_{las} is the power of the emitted laser light, S is the illuminated target area, R is the distance between the lidar and target, τ represents the optical propagation through the **O1** and **O2** optical systems, ε the target reaction on the incident laser light, and γ the geometric form-factor for the propagation of the laser light and the response signal through the ambient media (air, water etc.). This simple equation covers the majority of known areas of laser remote sensing. Some of these applications we discuss below.

LIDAR APPLICATIONS

There are three general spheres which are investigated by laser remote sensing: air, water and land. The modern lidars used for such investigations are mounted on board a car, boat, aircraft, helicopter, or a space-shuttle. The well-known police laser "speed-gun" is an example of a simple lidar device. Another simple application of the lidar technique is the device for measurement of distance; pulsed lasers are employed in such devices. In this case it is easy to show that the function γ (Eq.1) would contain a multiplier roughly equal to $\varphi(t-2R/c)$, where the function $\varphi(t)$ describes the time profile of the emitted laser pulse, R is the distance, and c the speed of light. The detected pulse reflected from the sounding target is time-shifted relative to the original laser pulse by $t = 2R/c$. Thus, the distance between the lidar and the target is $R = \frac{1}{2}ct$. The most famous application of this technique was the precise measurement of the distance between Earth and the Moon.

The example discussed above discloses the basic problem of laser remote sensing: we have to find some calibration procedure based on the "internal" parameters of the detected signal if we want to conduct quantitative measurements. It is impossible to calibrate a lidar in the same way as a common spectroscopic device like a spectrofluorimeter because the conditions γ of light propagation and interaction with the target ε cannot be stabilised. Thus, we need some internal calibration standard, and all remote measurements should employ the ratio (or difference) of at least two signals, one of them a calibration signal. In the latter example we used the relative timeshift between the "shot" and detected pulses.

The calibration standard can be of various kinds: (a) the precise time of the laser shot (see above), or the time interval between two events; (b) quasi-constant signals such as the water Raman signal (for sounding in aqueous media) or nitrogen Raman signal (for atmospheric sounding); (c) the laser light frequency (Doppler sounding) etc. The examples below demonstrate the application of the principle of relative measurements.

Another problem is solar blindness, i.e. the requirement for independence of lidar sensing on illumination by the sun. This problem could be solved by the use of spectral discrimination and pulse gating of the detector synchronised with the emitted laser pulse. In the gated systems the detector is 'open' only for a short period during the approach of the response signal pulse. Mostly the useful return signal can be detected within the 'open' period (10^{-6} - 10^{-9} sec). The number of solar photons which pass through the time and spectral gates of the detector optics must be less than those for a response signal from the target. The custom devices for spectral discrimination are optical filters and spectrometers (monochromator etc.).

The technique designated by the acronym DIAL (DIfferential Absorption LIDAR) is a good demonstration of the principle of relative measurements. Let us suppose that we have to detect some substances having a specific absorption band (see Fig. 3). We can probe it by two laser

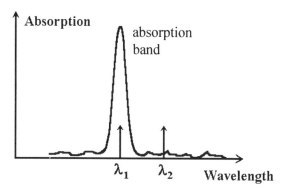

Figure 3. Fictional absorption spectra of detected substances.

wavelengths λ_1 and λ_2, where λ_1 is positioned in the absorption peak, while λ_2 is positioned outside it and is not influenced by the presence of these substances. In this case Eq.1 yields

$$\Phi = \frac{P_{det}(\lambda_1)}{P_{det}(\lambda_2)} \cong Cn_{sub}$$ because both detected signals contain the same depen-dence on P_{las}, S,

R, τ and γ (see Eq. 1), which vanish in their ratio, and absorption of the laser light at the wavelength λ_1 is thus proportional to the concentration of the measured substances. In the latter expression C is a constant and n_{sub} is a concentration of the substances presented in the parameter ε (Eq. 1).

REMOTE SENSING OF AIR

The major objects of remote atmospheric sensing are various gases, aerosols, water vapour (clouds), ice crystals, temperature, pressure etc. Aerosols or clouds can serve as tracers of the atmosphere dynamics, for example, for wind rate and direction. The lower layers of the atmosphere (up to tens of kilometres) could be monitored from land- or aircraft-based lidars, while the stratosphere can be sounded from the space-shuttle (Hinkley, 1976; Zuev and Romanovskii, 1990; Fishman, 1991; Couch et al., 1991; Kugeiko and Malevich, 1991; McCormick et al., 1993) Below we consider some particular cases of the remote atmospheric sensing.

Measurement of Extinction Coefficient

During its propagation through a medium (i.e. the atmosphere) having an extinction coefficient α, the absorption and scattering of the laser beam and response signals are reflected in the function γ (Eq.1) as $\exp(-\alpha ct)$ with ranging distance $R = \frac{1}{2}ct$. Thus, the extinction coefficient

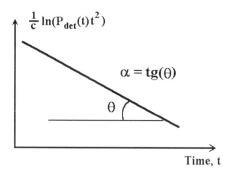

Figure 4. The modified kinetics of return signal P(t).

(Kunz and Deleeuw, 1993) could be determined as a tangent of the function $c^{-1}\ln(P_{det}(t)t^2)$, see Fig. 4. This method also employs the principle of relative measurements because the logarithm procedure is equivalent to the determination of ratio.

Detection of Aerosols

Aerosols, for example particulate pollutants, as well as water vapour and ice crystals (in clouds) can be detected by employing the Rayleigh and Mie scattering of the incident laser light (Measures, 1984). This is elastic scattering, i.e. the scattered signal has the same wavelength as the laser light, and its intensity, scattering diagram and polarisation properties are dependent on the aerosol concentration, the size of the particles and their refractive indices. One can recognise the presence of some aerosols and measure their spatial distribution by analysing the intensity and polarisation properties of the backscattered signal (Measures, 1984). The principle of relative measurement demands the use of some calibration signal. Commonly the N_2 Raman signal is employed as a calibration standard. The concentration of nitrogen in the atmosphere is assumed to be a quasi-stable.

Thus, there are two signals measured simultaneously (at different wavelengths): (a) the backscattered signal $P_{det}(\lambda_{las})$, and (b) the nitrogen Raman signal $P_{det}(\lambda_{N_2})$. Equation 1 yields the following formula for an aerosol concentration:

$$n_{aerosol} \sim \Phi = \frac{P_{det}(\lambda_{las})}{P_{det}(\lambda_{N_2})} \quad .$$

A good example is afforded by the eruption in June 1991 of the volcano Pinatube, which pumped a tremendous amount of aerosols into the troposphere and stratosphere. Since then many lidar stations around the world have observed the result of this eruption during the years 1991-94.

Raman LIDAR

In contrast to Rayleigh-Mie scattering, Raman scattering of laser light provides detailed information about the nature of the scatterer (Inaba and Kobayasi, 1969; Measures, 1984). The Raman signal is spectrally shifted relative to the fundamental frequency of the incident light by the frequency of molecular vibrations: $v_R^{s,as} = v_{las} \pm v_{mol.}$; $v_R^s = v_{las} - v_{mol.}$ is called the Stokes

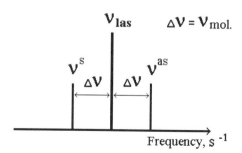

Figure 5. The frequencies of Raman scattering.

component, and $v_R^{as} = v_{las} + v_{mol.}$ the anti-Stokes component (see Fig.5). Commonly the Stokes component is used in lidar sensing as it is more intense. Each kind of molecule has its own specific spectrum of Raman components v_{mol} and molecules can be recognised by these components. Raman lidar measures the intensity of the return signal for the Stokes component of the searched molecules and also the Stokes component of the N_2 molecules for internal calibration. Thus, equation 1 yields the following formula for the concentration of the detected molecules:

$$n_{mol.} \sim \Phi = \frac{P_{det}(\lambda_{mol.}^R)}{P_{det}(\lambda_{N_2}^R)} \quad , \qquad \lambda = \frac{c}{v} \quad .$$

Raman lidars are widely used for detection of various gases such as CO_2, O_3 (ozone), SO_2, H_2O (water vapour) etc.

Fluorescence LIDAR

Many molecules, and in particular a number of aromatic molecules, fuel and other oil components, can fluoresce under light excitation. Fluorescence is the process of depopulating the molecular excited states via the emission of light quanta. Different types of molecules have different types of fluorescence spectra, and this feature is used for their detection and identification. The position and bandwidth of the fluorescence spectrum are stable and red-shifted relative to the wavelength of the excitation light. The power density of the emitted fluorescence is given by $P_{fluor.} = \dfrac{1}{4\pi^2}\sigma_{abs}\eta P_{illum.}$, where $P_{illum.}$ is the power density of the incident light, σ_{abs} is the absorption cross-section, and η is the fluorescence quantum yield. Actually the latter formula is the expression for the ε function in Eq. 1. Thus, we can find the following formula for the normalised detected signal of the fluorescent target:

$$n_{fluor.} = C\Phi = C\frac{P_{det}(\lambda_{fluor})}{P_{det}(\lambda_{N_2}^R)} \quad .$$

Here the constant C includes the parameters σ_{abs} and η.

Doppler LIDAR

Doppler lidar serves for measurement of the speed of aerosols or clouds (Billbro, 1980; Akhmanov et al., 1991; Korb et al., 1992). This method uses the well-known Doppler phenomenon whereby the fundamental frequency of the incident laser light v_{las} is shifted by $\Delta v = (2v/c)v_{las}$, where v is the component of the aerosol motion in the direction of the laser beam and c is the speed of light. The Doppler signal is created by moving aerosols or clouds. From the equation for the frequency shift Δv it is clear that the infrared lasers (CO_2, Nd:YAG etc.) having a lower fundamental frequency v_{las} should be used to achieve the best accuracy. In this spectral range it is easier to discriminate the fundamental and backscattered Doppler signals. Moreover, the laser must be frequency-stabilised. Various methods can be used for such discrimination, for example Fabry-Perot interferometry, edge filter techniques etc.

Measurement of Temperature

There are a few methods for remote measurements of air temperature. Perhaps one of the most interesting and reliable is the method based on use of the rotational Raman scattering by O_2 and/or N_2 molecules (Vaughan et al., 1993). This method is self-calibrated and based on ratio measurements of the rotational Raman components. The population of these components is dependent on the local air temperature. In section "Raman LIDAR" we have considered the case of vibrational Raman scattering; it is known that the fine structure of each vibrational band includes rotational components, which can be resolved by modern spectroscopic techniques.

It was shown by Vaughan et al. (1993) that the "target reaction" (see Eq.1) for the formation of the rotational Raman components equals:

$$\varepsilon(J,T,...) = F(J,...)\frac{1}{kT}\exp\left(-\frac{hcB_0 J(J+1)}{kT}\right) \quad ,$$

where F is a function whose exact form need not here concern us, and the other parameters are: J - line number, B_0 - rotational constant, h - Planck's constant, k - Boltzmann constant, T - absolute temperature. Thus the ratio of two detected components (J_1 and J_2) is given by:

$$\Phi(T) = \tilde{C} \exp\left(-\frac{hcB_0[J_1(J_1+1) - J_2(J_2+1)]}{kT}\right) \quad,$$

or if $J_2 = J_1-1$, i.e. if two neighbouring lines are used, then

$$\Phi(T) = \tilde{C} \exp\left(-\frac{2hcB_0J_1}{kT}\right) \quad,$$

These formulae allow one to measure the air temperature.

REMOTE SENSING OF WATER

The major objects in the remote sensing of water are depth (bathymetry), pollution, biological productivity (phytoplankton), aqueous extinction coefficient, temperature, salinity etc. The same principles and techniques of lidar sensing are employed in water control as in the case

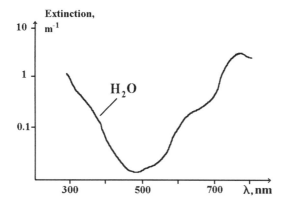

Figure 6. Extinction coefficient of distilled water.

of atmospheric lidar sensing. The main difference lies in the timescale and in the conditions of light propagation through the significantly different media of air and water. Aqueous media have a larger extinction coefficient than air, and that causes a decrease of the light propagation length to 1 - 100 m, depending on the wavelength of light involved. The variation of the extinction coefficient of distilled water with wavelength is presented in Fig.6 (Smith and Tyler, 1976). Consequently the lidar timescale 'shrinks' to the nanosecond range, in contrast to the microsecond range employed in air remote sensing. Thus, faster electronics must be used in sounding aqueous media and only subsurface waters can be sounded.

Bathymetry (Depth Measurement)

The principle of depth measurement is very simple. One measures the time between two reflections of the incident laser pulse: from the surface of the water (t_1) and from the bottom (t_2) respectively (see Fig.7). The water depth can be determined from the formula $z = \frac{1}{2}(c'\Delta t)$, where c' is the speed of light in the water. This method works effectively for shallow shelves (0-100 m)

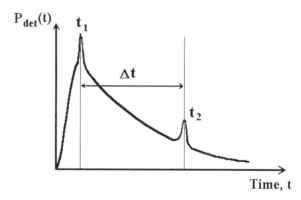

Figure 7. Typical kinetics of return echo-signal.

with clear water (Hickman and Hogg, 1969; Kim, 1977; Muirhead and Cracknell, 1986; Steinvall, 1993).

Extinction Coefficient of Water

The method of remotely measuring the extinction coefficient of water is the same as for air measurements (see paragraph "Measurement of Extinction Coefficient"). The only difference is in the timescale of the return signals. This method was used in various works (Kim, 1973; Zakharov and Goldin, 1984; Bukin et al., 1990).

Detection of Phytoplankton

Measurement of Chlorophyll Concentration. Phytoplankton is one of the most important species in natural waters. It is a form of algae, which conducts the process of photosynthesis and serves as a food for zooplankton, which in turn is a food for fish. Thus, phytoplankton is a basis of life in the ocean, and it determines the biological productivity of natural waters.

These algae contain various photosynthetic pigments and the most important is chlorophyll a (Chl a). Upon the excitation of phytoplankton by laser light the energy of the absorbed photons

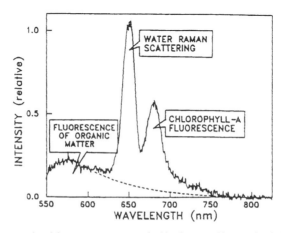

Figure 8. Typical response signal from ocean water excited by the second harmonic of a Nd:YAG-laser (λ_{las} = 532 nm) (Chekalyuk et al., 1992).

140

transfers through the photosynthetic molecular complexes containing chlorophyll and other pigments to the reaction centres (RC). In the RC excitation is trapped and used for the biochemical stage of photosynthesis, which in particular produces oxygen (O_2). Such reaction centres could be considered as traps for excitations randomly hopping among the photosynthetic pigments in the light-harvesting antennae of the algae. A portion of the absorbed light energy could however be emitted by the chlorophyll molecules. The corresponding fluorescence signal, having a maximum at wavelength λ=685-690 nm, contains information on the concentration of chlorophyll and its fluorescence quantum yield in the light-harvesting antennae. It is clear that in general the fluorescence quantum yield depends on the status of the RCs (Chekalyuk et al., 1992; Chekalyuk and Gorbunov, 1992). Thus the quantum yield provides information on the photosynthetic activity of intact algae (see next paragraph). The intensive research in the field of lidar remote sensing of phytoplankton began in the 70's (Kim, 1973; Mumola et al., 1973; Klyshko and Fadeev, 1978; Fadeev et al., 1980a). In 1975 Fadeev et al. (1980a) introduced for the first time the idea of calibrating the phytoplankton fluorescence by a water Raman signal having Stokes shift $\Delta v = 3440$ cm^{-1} (see Fig. 8). In 1978 this method was theoretically established by Klyshko and Fadeev (1978), and it subsequently found wide usage in the lidar sounding of phytoplankton (Fadeev et al., 1980b; Demidov et al., 1981; Hoge and Swift, 1981; Demidov et al., 1988).

According to work by Klyshko and Fadeev (1978) the ratio of the phytoplankton fluorescence $P_{det}(\lambda_{Chl})$ and the water Raman signal $P_{det}(\lambda_{H_2O})$, called the "fluorescence parameter" (Klyshko and Fadeev, 1978; Fadeev et al., 1980), is proportional to the Chl a concentration: $n_{Chl\ a} = C\Phi = CP_{det}(\lambda_{Chl\ a})/P_{det}(\lambda_{H_2O})$. Here C is a coefficient dependent on the spectroscopic parameters of the phytoplankton pigments, the efficiency of energy transfer in the algae light-harvesting antennae and the functional status of the RC. In a series of experimental studies (Fadeev et al., 1980b; Demidov et al., 1981; Demidov et al., 1988; Vedernikov et al., 1990) it was revealed that under certain conditions of illumination (appropriate power density of excitation etc.) this coefficient has the value C = 2.6 ± 0.3 μg litre^{-1}. Thus, the method could be used in mesotrophic and eutrophic waters, i.e. waters with medium or high levels of biological productivity.

Evaluation of Phytoplankton Photosynthetic Activity. Because the parameter C (see previous paragraph) is linearly proportional to the Chl a fluorescence quantum yield, it provides

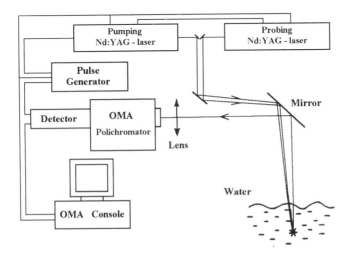

Figure 9. Pump-probe lidar scheme (Chekalyuk et al., 1992).

an opportunity for remote measurement of the photosynthetic activity of phytoplankton. Recently Chekalyuk et al. (1992) and Chekalyuk and Gorbunov (1992) introduced the method of double-pulse lidar sensing for the remote determination of phytoplankton photosynthetic activity. The scheme of double-pulse lidar is presented in Fig. 9. The idea of this method is quite straightforward. One must measure the fluorescence parameter Φ twice: (a) after a common single-pulse excitation (Φ_1), when only the 'probe' laser is involved; and then (b) after a double-pulse excitation (Φ_2), when fluorescence is detected only from a weak 'probe' pulse following 40 - 70 μs after the intense 'pump' pulse. The pump pulse causes photodynamic closure of the RCs and the 'probe' pulse excites the photosynthetic system having a higher fluorescence quantum yield because the traps of excitation (i.e. the RCs) are blocked. Thus, the ratio $\xi = \Phi_2/\Phi_1$ contains information about the number of initially opened and closed RCs, i.e. about the ability of algae to photosynthesise (photosynthetic activity).

LIDAR Determination of Depth Distribution of Phytoplankton

In the theoretical works by Demidov and Fadeev (1980, 1983) it was shown that the principle of internal calibration by the water Raman signal could be employed in kinetics measurements of the fine structure of the depth distribution of phytoplankton Chl a. Following Demidov and Fadeev (1980) one can write the following expression for the determination of this distribution:

$$ n_{Chl\,a}(z) = \hat{C}\,\frac{P_{det}(\lambda_{Chl\,a}, z)}{P_{det}(\lambda_{H_2O}, z)}\,e^{\Delta k z} \quad , $$

where $z = \frac{1}{2}(c\,t)$ is the depth of the water layer probed, and $\Delta k = k(\lambda_{Chl\,a}) - k(\lambda_{H_2O})$ is the difference between the optical absorption coefficients by water at the wavelengths of Chl a fluorescence and of the water Raman signal.

Detection of Oil and Dissolved Organic Matter

Oil is one of the most common pollutants of ocean and fresh waters. Remote sensing of oil spills and dispersed oil is very important for environmental control. In some cases dissolved organic matter could be considered a pollutant (or a result of pollution) as well, but it is common to find dissolved organic matter (DOM) in any waters, including the cleanest waters of the Indian and other oceans.

The method of detecting oil (or fuel) and DOM in an aqueous environment is based on their fluorescence properties (O'Neil et al., 1980; Bristow et al., 1981; Burlamacchi et al., 1983; Abroskin et al., 1988; Hengstermann and Reuter, 1990; Patsaeva et al., 1991; Filippova et al., 1993). Thus, we have a case of fluorescence lidar (see above). Both types of pollutant have broad overlapping spectra (tens of nm) of fluorescence in the blue-green-yellow range of the spectrum. Thus, commonly UV laser sources like the N_2-laser or excimer lasers are used for their detection. The water Raman signal routinely serves as an internal calibration signal. When both DOM and oil are present in the water, it is a problem to distinguish between them because their spectra have a relatively similar structure and they are strongly overlapped. However, there is some difference; moreover different oils have slightly different, but still broad spectra. In addition, these species have different fluorescence lifetimes (Patsaeva et al., 1991; Schade and Bublitz, 1992; Filippova et al., 1993). All these properties are employed in the lidar sensing of oil and DOM. These lidars use multiwavelength excitation and detection and/or kinetics measurement of the response signal (Schade and Bublitz, 1992; Filippova et al., 1993).

Remote Sensing of Temperature and Salinity

In several works (Leonard and Chang, 1974; Leonard et al., 1979; Bekkiev et al., 1983; Gogolinskaya et al., 1986) it has been shown that the Raman signal of water can be used for remote determination of water temperature and salinity. In general, the Raman signal is quite stable, but nevertheless its fine structure is sensitive to both temperature and salt content. This dependence for water samples was first reported by Leonard and Chang (1974), and then employed in lidar sensing (Leonard et al., 1979; Bekkiev et al., 1983; Gogolinskaya et al., 1986). Fig. 10 schematically represents the general dependence of the main Stokes band (Δv = 3100-3700 cm^{-1}) of the water Raman signal on the temperature. One can see that the intensities of the Raman signal at wavelengths λ_1 and λ_2 are differently dependent on temperature; thus, their ratio

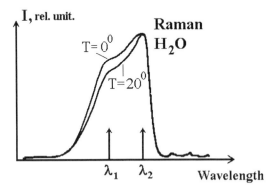

Figure 10. Raman spectrum of water.

can be used for the evaluation of temperature. In liquid water there are at least two forms: monomeric and dimeric. The monomeric form is responsible for the λ_2 component, the dimeric form for the λ_1. The equilibrium between these forms depends on the temperature. If the water contains some salt, then it will also influence this equilibrium, though the functionality of its influence on the Raman band is somewhat different.

More accurate measurements involve four or even more wavelengths. In general the shape of the Raman signal (G) is a function of at least two parameters: temperature (T) and salinity (S)-

$$G = G(\lambda, T, S) = G_0(\lambda) + \frac{dG}{dT}\Delta T + \frac{dG}{dS}\Delta S + ... \cong G_0(\lambda) + C_1(\lambda)(T - T_0) + C_2(\lambda)(S - S_0).$$

This formula establishes the theoretical basis for the simultaneous measurement of water temperature and salinity. Best results can be achieved when the whole Raman spectrum is involved in the calculations (Gogolinskaya et al., 1986): the accuracy of the temperature and salinity determinations are about 0.01^0 C and a few promiles. (For reference the salinity of ocean waters is typically about 34 promiles).

REMOTE SENSING OF LAND VEGETATION

In this part we consider only the lidar sensing of vegetation. The fluorescence of land flora is widely used in this sensing. The leaves of trees or grass contain a very complicated photosynthetic apparatus similar to the photosynthetic apparatus of phytoplankton, and all the problems discussed in section "Detection of Phytoplankton" have a place in the case of land vegetation. The photosynthetic apparatus of green plants contains dozens of photosynthetic pigments (chlorophyll *a*, *b*, *c*; carotenoids etc.), the photosynthetic reaction centres contain specific molecules like pheophytin, quinones etc., and there are other molecules like vitamin K_1 and so on. Moreover chlorophylls and proteins can create chlorophyll-protein associates with different spectroscopic properties. The latter is the reason why the common Chl *a* displays several spectral forms with fluorescence maxima ranging from 670 nm to 740 nm (Govindjee, 1982).

All pigments, pigment-proteins, and RC complexes are associated with two major light-harvesting systems: photosystem I (PSI) and photosystem II (PSII). Both photosystems contain major photosynthetic pigments, including Chl *a*, and reaction centres, but they have different fluorescence properties. Thus fluorescent quanta emitted from the Chl molecules associated with the PSI are red shifted (λ_{fl} = 720-740 nm) relative to the fluorescence of PSII (λ_{fl} = 685-690 nm). In addition, upon excitation in the UV band (Govindjee, 1982; Chappele et al., 1984a,b; Chappele et al., 1985; Chappele et al., 1989; Cecchi et al., 1986; Broglia, 1993), the fluorescence of pigments other than Chl is observed at wavelengths shorter than 685 nm. Both photosystems are employed in the process of photosynthesis and reflect the biological condition of plants, grasses, crops etc., under the influence of environmental stress.

Vegetation sounding lidar has two major areas of survey application: (a) the identification of plants and the study of their stress under environmental impact, and (b) the estimation of tree-heights and consequently the volume and/or green weight of wood on the ground.

Sensing of Vegetation Characteristics

It was disclosed in various studies (Chappele et al., 1984a,b; Chappele et al., 1985; Chappele et al., 1989; Cecchi and Pantani, 1991; Broglia, 1993) that the fluorescence spectra of different vegetation have different shape. Moreover, this shape depends on the photosynthetic

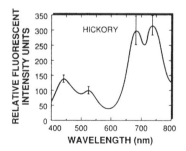 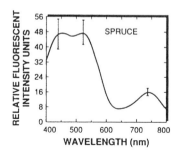

Figure 11. Fluorescence spectra of two plant types excited by N_2-laser (Chappele et al., 1989).; reproduced by permission of Laser Focus World.

activity of these species and environmental stress factors such as 'acid rain', the presence of heavy metals, nutrient stress, drought etc. (Chappele et al., 1984a,b). Fig.11 presents some typical spectra of two plants obtained under UV excitation by a pulsed N_2-laser, λ_{las} = 337 nm, (Chappele et al., 1989). One can see that there are at least four wavelengths: 440, 525, 685 and 740 nm, which can characterise the investigated plants and serve for their identification. Commonly, the ratios of the laser-induced fluorescence (LIF) at these wavelengths are used for this purpose (the principle of relative measurements in lidar detection), for example, $P_{det}(\lambda = 440$ nm)/ $P_{det}(\lambda = 685$ nm). Thus, the basic idea of the detection of vegetation by a lidar technique is clear, and the major efforts of current research in this field are focused on collecting data and understanding the fundamental problems concerning the influence of environmental factors on the generation of response fluorescence. In turn, this calls for a proper understanding of how laser light interacts with such a complicated system as the photosynthetic apparatus of intact plants.

LIDAR Mensuration of Forest

Airborne lidar systems have been used to assess forest-canopy density, the height of trees, green biomass of forests etc. (Nelson et al., 1984; 1988; Ritchie et al., 1993, Weltz et al., 1994). These systems employ a method analogous to the bathymetry described above. So, a pulse lidar

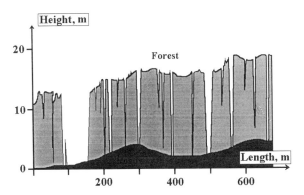

Figure 12. Profile of lidar sounding of forest.

system operating from on board an aircraft detects the return signals reflected both from the top of the canopy, and from the ground when the laser beam comes through forest gaps or between the trees. These profiles of the return signals (see Fig.12) allows measurement of tree heights: $h=\frac{1}{2}(\Delta ct)$, where Δt is the time between the return pulses reflected from the canopy top and the ground. The biomass could be evaluated from the area under the curve for the signal reflected from the canopy top (see the lighter grey area in Fig. 12). It was found that lidar data agree with ground-based estimation with fair accuracy, the discrepancy being only about 3% (Chappele et al., 1989).

REFERENCES

Abroskin, A.G., Nolde, S.E., Fadeev, V.V., and Chubarov, V.V., 1988, *Sov. Phys. Dokl.* (English translation from *Dokl. Acad. Nauk SSSR*, 1988, 299(2):351), 33:215.

Akhmanov, S.A., Gordienko, V.M., Kosovsky, L.A., Kurochkin, N.N., Pogosov, G.A., and Priezzhev, A.V., 1991, *Izvestiya Akad. Nauk SSSR, Seriya Fiz.* 55:194.

Bekkiev, A.I., Gogolinskaia, T.A., and Fadeev, V.V., 1983, *Sov. Phys. Dokl.* (English translation from *Dokl. Acad. Nauk SSSR*, 1983, 271(4):849), 28:639.

Bilbro, J.W., 1980, *Optical Engineering.* 19:533.

Bristow, M., Nielsen, D., Bundy, D., and Furtek, R., 1981, *Appl. Optics.* 20:2889.

Broglia, M., 1993, *Appl. Optics.* 32:334.

Bukin, O.A., Iliichev, V.I., Kritskii, I.A., Pavlov, A.N., 1990, *Dokl. Acad. Nauk SSSR*, 312:972.

Burlamacchi, P., Cecchi, G., Mazzinghi, P., and Pantani, L., 1983, *Appl. Optics.* 22:48.

Cecchi, G., and Pantani, L., 1991, *in:* "Physical Measurements and Signatures in Remote Sensing," 5th International Colloquium, 319 (Ch.164, v.2):687.

Chappelle, E.W., Wood, F.M., McMurtrey, J.E., and Newcomb, W., 1984a, *Appl. Optics.* 23:134.

Chappelle, E.W., McMurtrey, J.E., Wood, F.M., and Newcomb, W.W., 1984b, *Appl. Optics.* 23:139.

Chappelle, E.W., Wood, F.M., Newcomb, W.W., and McMurtrey, J.E., 1985, *Appl. Optics.* 24:74.

Chappelle, E.W., Williams, D.L., Nelson, R.F., and McMurtrey, J.E., 1989, *Laser Focus World.* 25 (6):123.

Chekalyuk, A.M., Demidov, A.A., Fadeev, V.V., and Gorbunov, M.Yu., 1992, *in*: "XVIIth Congress of International Society for Photogrammetry and Remote Sensing, International Archives of Photogrammetry and Remote Sensing," 29(B7): 878.

Chekalyuk, A.M., and Gorbunov, M.Yu., 1992, *in*: "XVIIth Congress of International Society for Photogrammetry and Remote Sensing, International Archives of Photogrammetry and Remote Sensing," 29(B7):897.

Couch, R.H., Rowland, C.W., Ellis, K.S., Blythe, M.P., Regan, C.P., Koch, M.R., Antill, C.W., Kitchen, W.L., Cox, J.W., Delorme, J.F., Crockett, S.K., Remus, R.W., Casas, J.C., and Hunt, W.H., 1991, *Opt. Enginer.* 30:88.

Demidov, A.A., Baulin E.V., Fadeev V.V., and Shur L.A., 1981, *Oceanology (Academy of Sciences of the USSR)* (English translation from *Oceanology*, 1981, 21(1):174), American Geophysical Union, 21:121.

Demidov, A.A., and Fadeev, V.V., 1980, *Sov. Phys. Dokl.* (English translation from *Dokl. Acad. Nauk SSSR*, 1980, 255(4):850), 25:1002.

Demidov, A.A., and Fadeev, V.V., 1983, *Sov. Phys. Dokl.* (English translation from *Dokl. Acad. Nauk SSSR*, 1983, 271(2):344), 28:574.

Demidov, A.A., Chekalyuk, A.M., Lapshenkova, T.V., and Fadeev, V.V, 1988, *Soviet Meteorology and Hydrology* (English translation from *Meteorologiya i Gidrologiya (USSR)*, 1988, 6:62), Allerton Press, Inc., 6:43.

Fadeev, V.V., Rubin, L.B., Kust, S.V., Petrosyan, V.S., Tunkin, V.G., and Haritonov, L.A., 1980, *in:* "Complex Investigations of Ocean Nature," The First Conference of Moscow State University on Problems of World Ocean, Moscow, December 1975, p.213.

Fadeev, V.V., Demidov, A.A., Klyshko, D.N., Koblentz-Mishke, O.I., and Fortus, V.M., 1980, *Annals of Institute of Oceanology (USSR)*, 90:219.

Filippova, E.M., Chubarov, V.V., and Fadeev, V.V., 1993, *Can. J. Appl. Spectr.* 38:139.

Fishman, J. 1991, *Env. Science & Technol.* 25:612.

Hengstermann, T., and Reuter, R. 1990, *Appl. Optics.* 29:3218.

Hickman, G.D., and Hogg, J.E., 1969, *Remote Sensing of Environment* 1:47.

Hinkley, E.D., 1976, *Optical and Quantum Electronics* 8:155.

Hoge, F.E., and Swift, R.N., 1981, *Appl. Optics.* 20:3197.

Gogolinskaia, T.A., Patsaeva, S.V., and Fadeev, V.V., 1986 *Sov. Phys. Dokl.* (English translation from *Dokl. Acad. Nauk SSSR*, 1986, 290(5):1099) 31:820.

Govindjee (ed.), 1982, "Photosynthesis," Acad. Press, New York, London.

Inaba, H., and Kobayasi, T., 1969, *Nature*, 224:170.

Kim, H.H., 1977, *Appl. Optics.* 16:46.

Kim, H.H., 1973, *Appl. Optics.* 12:1454.

Klyshko, D.N., and Fadeev, V.V., 1978 *Sov. Phys. Dokl.* (English translation from *Dokl. Acad. Nauk SSSR*, 1978, 238(2):320), 23:55.

Korb, C.L., Gentry, B.M., and Weng, C.Y., 1992, *Appl. Optics.* 31:4202.

Kugeiko, M.M., and Malevich, I.A., 1991, *Soviet Journal of Remote Sensing*, 9:69.

Kunz, G.J., and Deleeuw, G., 1993 *Appl. Optics.* 32:3249.

Leonard, D.A., and Chang, C.H., 1974, *United States Patent: 3,806,727.* Apr.23.

Leonard, D.A., Caputo, B., and Hoge F.E. 1979, *Appl. Optics.* 18:1732.

Ligda, M.G.H., 1963, *in*: "Proc. Conf. Laser Technol.," 1st, San Diego, Calif., p.63.

McCorMick, M.P., Winker, D.M., Browell, E.V., Coakley, J.A., Gardner, C.S., Hoff, R.M., Kent, G.S., Melfi, S.H., Menzies, R.T., Platt, C.M.R., Randall, D.A., and Reagan, J.A., 1993, *Bullet. Amer. Meteor. Soc.* 74:205.

Measures, R.M., 1984, "Laser Remote Sensing," Wiley, New York, Chichester.

Muirhead, K., and Cracknell, A.P., 1986, *Int. J. Rem. Sensing.* 7:597.

Mumola, P.B., Jarrett, O., and Brown, C.A., 1973, *in:* "NASA conference on the use of lasers for hydrographic studies," NASA SP-375, p.137.

Nelson, R., Krabill, W., and Maclean, G., 1984, *Remote Sens. Env.* 15:201.

Nelson, R., Krabill, W., and Tonelli, J., 1988, *Remote Sens. Env.* 24:247.

O'Neil, R.A., Buje-Bijunas, L., and Rayner, D.M., 1980, *Appl. Optics*, 19:863.

Patsaeva, S.V., Fadeev, V.V., Filippova, E.M., and Chubarob, V.V., 1991, *in:* "12th Asian Conference on Remote Sensing, Singapore," p.Q-12-1.

Ritchie, J.C., Evans, D.L., Jacobs, D., Everitt, J.H., and Weltz, M.A., 1993, *Transactions of the ASAE* 36:1235.

Schade, W., and Bublitz, J., 1992, *Instit. of Physics Confer. Ser.* 128:317.

Smith, R.C., and Tyler J.E., 1976, *Photochem. Photobiol. Rev.* 1:117.

Steinvall, K.O., 1993, *Optical engineering.* 31:1307.

Vaughan, G., Wareing, D.P., Pepler, S.J., Thomas, L., and Mitev, V., 1993, *Appl. Optics.* 32:2758.

Vedernikov, V.I., Vshyvcev, V.S., Demidov, A.A., Pogosian, S.I., Sukhanova, I.N., Fadeev, V.V., and Chekaluk, A.M., 1990, *Oceanology (Academy of Sciences of the USSR)* (English translation from *Oceanology*, 1990, 30:848), American Geophysical Union, 30:628.

Weltz, M.A., Ritche, J.C., and Fox, H.D., 1994, *Water Resources Research.* 30:1311.

Zakharov, A.K., and Goldin, Y.A., 1984, *Izvestiya Akad. Nauk SSSR, Seriya Fiz. Atmosphery I Okeana* 20:661.

Zuev, V.V., and Romanovskii, O.A., 1990, *Soviet Journal of Remote Sensing* 6:80.

NONLINEAR OPTICS

Ravi Singhal

Department of Physics & Astronomy
University of Glasgow
Glasgow G12 8QQ, Scotland

INTRODUCTION

Nonlinear optical effects were first demonstrated by Franken and colleagues in 1961[1]. The 694.3 nm light from a ruby laser was focused on a quartz crystal and an extremely weak beam of wavelength 347.15 nm was observed. Even though the efficiency of frequency-doubling was only about 10^{-8}, an experimental demonstration of the nonlinear effect was highly significant for generating enthusiasm in the scientific community.

Over the following thirty years, the development of practical devices that use nonlinear optical effects has been such that they now totally dominate the electro-optical technologies. The availability of intense laser radiation was important because the large electromagnetic fields associated with lasers are normally required for observing nonlinear effects. When light passes through materials, the constituents of the material are polarised by the electric field of the incident light wave. For moderate light intensities, the induced polarisation P, essentially representing the changes in the separation of positive and negative charges in individual atoms or molecules of the medium, is linearly proportional to the incident electric field E. The oscillating electric field induces an oscillating dipole moment which in turn radiates at the same frequency. However, for the light intensities readily available in pulsed lasers, electric fields are large enough to induce a polarisation that is no longer proportional to E, and nonlinear phenomena may manifest. The electric field and light intensity are related by $I = \frac{1}{2}\varepsilon_0 cE^2$, where ε_0 is the permittivity of the vacuum and c is the speed of light. This is equivalent to $E/\mathrm{V\,cm^{-1}} = 27.45\sqrt{I/\mathrm{W\,cm^{-2}}}$. For $I \sim 10^9\,\mathrm{W\,cm^{-2}}$ the electric field strength $E \sim 10^6\,\mathrm{V\,cm^{-1}}$ and is comparable to atomic field strengths. In the induced polarisation, nonlinear terms proportional to second and third orders of the electric field would be expected to be present and give observable effects for these high laser intensities.

---

[1] Raman, Pockels and Kerr effects were already known. Nonlinear optics as a separate field only developed after lasers became available.

An Introduction to Laser Spectroscopy
Edited by David L. Andrews and Andrey A. Demidov, Plenum Press, New York, 1995

The nonlinear medium has dimensions typically of a few tens of millimetres while the laser wavelengths are smaller by a factor of a thousand or more. The medium thus extends over many wavelengths and the interaction of the nonlinear polarisation with the laser beam, and subsequent build-up of the nonlinear effects, depends on proper phase matching of the incident wave and the induced polarisation. Lack of such phase matching was responsible for the low efficiency of frequency doubling in Franken's pioneering demonstration. With due consideration to this criterion, it is now possible to obtain frequency doubling efficiencies of greater than 70%.

Early research in nonlinear optics had concentrated on second harmonic generation and optical parametric oscillation to obtain continuously tunable coherent radiation in the wavelength range from about 200 nm to 2000 nm. The present scope of the field extends far beyond this and includes diverse subjects such as optical rectification, sum- and difference-frequency mixing, the Pockels and Kerr electro-optical effects, stimulated Raman, Brillouin and Rayleigh scattering, multiphoton absorption and ionisation, third- and higher-order harmonic generation, phase conjugation, self-focusing, self-phase modulation, X-ray lasers, high-temperature high-density plasma production, ultra-fast optical switching etc. The list appears to be endless!

NONLINEAR EFFECTS ON WAVE PROPAGATION

The instantaneous polarisation $\mathbf{P}$ of a medium in the presence of an electric field $\mathbf{E}$ can be expressed as a power series in $\mathbf{E}$

$$\mathbf{P} = \mu + \varepsilon_0 \chi^{(1)} \mathbf{E} + \varepsilon_0 (\chi^{(2)} \mathbf{E}^2 + \varepsilon_0 \chi^{(3)} \mathbf{E}^3 + \ldots\ldots) = \mathbf{P}_{\mathrm{L}} + \mathbf{P}_{\mathrm{NL}}, \tag{1}$$

where μ is the permanent dipole moment of the material along $\mathbf{E}$, and $\chi^{(1)}$ is the linear optical susceptibility related to the index of refraction n by $\chi^{(1)} = n^2 - 1 = \varepsilon/\varepsilon_0 - 1$, ε being the permittivity of the medium. The second- and third-order susceptibilities $\chi^{(2)}$ and $\chi^{(3)}$ are responsible for nonlinear phenomena of the respective order. The effect of $\mathbf{P}_{\mathrm{NL}}$ on the propagation of an electromagnetic wave in the medium may be studied by solving Maxwell's equations. These are

$$\nabla \times \mathbf{E} = -\frac{\partial}{\partial t}(\mu_0 \mathbf{H}) ; \quad \nabla \times \mathbf{H} = \sigma \mathbf{E} + \frac{\partial \mathbf{D}}{\partial t} ; \quad \nabla . \mathbf{E} = 0 , \quad \text{with}$$

$$\mathbf{D} = \varepsilon_0 \mathbf{E} + \mathbf{P} = \varepsilon_0 \mathbf{E} + \mathbf{P}_{\mathrm{L}} + \mathbf{P}_{\mathrm{NL}} . \tag{2}$$

On taking the curl of the first Maxwell equation and using a vector identity[2], we obtain

$$\nabla^2 \mathbf{E} = \mu_0 \sigma \frac{\partial \mathbf{E}}{\partial t} + \mu_0 \varepsilon \frac{\partial^2 \mathbf{E}}{\partial t^2} + \mu_0 \frac{\partial^2 \mathbf{P}_{\mathrm{NL}}}{\partial t^2} . \tag{3}$$

If $\mathbf{P}_{\mathrm{NL}}$ is zero then we obtain the equation which describes the propagation of electromagnetic waves in a linear medium. For nonzero values of $\mathbf{P}_{\mathrm{NL}}$ terms responsible for nonlinear optical phenomena are present. Thus far we have assumed that the nonlinear

---

[2] $\nabla \times \nabla \times \mathbf{A} = \nabla \nabla . \mathbf{A} - \nabla^2 \mathbf{A}$

polarisation is induced by the incident light wave. This need not be the case. For example, nonlinear polarisation may be caused by a second light wave, thus affecting the propagation of the first light field. Alternatively, nonlinear polarisation might be induced by the application of an external d.c. or a low frequency a.c. electric field. At present, the most widely used nonlinear processes are those associated with the second or third order susceptibility.

Let us consider the situation of three plane waves propagating in the z-direction with frequencies ω_i and amplitudes E_i with i = 1, 2 or 3, and restrict the nonlinear polarisation to second order. The nonlinear polarisation may be expressed in terms of frequency components oscillating at $\omega_k + \omega_l$ with amplitudes $\varepsilon_0 \chi^{(2)} E_k E_l$. The indices k and l may take integer values from -3 to +3; also $\omega_{-k} = -\omega_k$ and $E_{-k} = E_k^*$. Of the many possibilities expressed by $\omega_k + \omega_l$, we are normally interested only in the build up of a particular frequency component. The equation describing the evolution of this component may be isolated by using orthogonality properties of the exponential functions. For the particular case when $\omega_3 = \omega_1 + \omega_2$, we obtain (Yariv, 1985; Byer, 1977), for a lossless medium

$$\frac{dE_1}{dz} = \frac{i\omega_1 d_{eff} E_3^* E_2}{n_1 c} e^{-i(k_3-k_2-k_1)z} \quad ,$$

$$\frac{dE_2}{dz} = \frac{i\omega_2 d_{eff} E_1 E_3^*}{n_2 c} e^{-i(k_1-k_3+k_2)z} \quad , \tag{4}$$

$$\frac{dE_3}{dz} = \frac{i\omega_3 d_{eff} E_1 E_2}{n_3 c} e^{-i(k_1+k_2-k_3)z} \quad .$$

where n_i are the indices of refraction and $k_i = 2\pi/\lambda_i = n_i \omega_i/c$. The constant d_{eff} is related to the second order nonlinear susceptibility of the medium and is listed in tables (Byer, 1977) for a number of commonly used materials. Equations (4) are the evolution equations and with them we are in a position to analyse a variety of second-order nonlinear processes.

In a medium, a range of nonlinear effects is possible. However, of the many possibilities, normally only one effect dominates and is the one for which the energy conservation and phase matching conditions are satisfied simultaneously. In the above example, if we consider ω_1 and ω_2 are the frequencies of the input photons and ω_3 the frequency of the output photon then $\omega_3 = \omega_1 + \omega_2$ simply represents the energy conservation condition and corresponds to the particular case of sum-frequency generation. Similarly, $\omega_2 = \omega_3 - \omega_1$ represents the case of difference-frequency generation where ω_3 and ω_1 are the input and photons of frequency ω_2 are generated by the nonlinear interaction. In all cases, however, coherent energy transfer must occur between the propagating fields and the nonlinear polarisation, and this is expressed by the phase-matching condition $\mathbf{k}_3 = \mathbf{k}_1 + \mathbf{k}_2$ which may also be seen to represent the conservation of photon momentum. The particular case of phase-matching in second harmonic generation (SHG) will be discussed in detail in a later section.

Symmetry Constraints on $\chi^{(2m)}$

Although the induction of odd order polarisation in a medium is not subject to any symmetry constraint, even order susceptibilities vanish for materials which possess inversion symmetry. This precludes the observation of second order nonlinear effects in gaseous media and in centrosymmetric crystals. Centrosymmetric crystals are those whose structure remains unchanged upon inversion i.e. on replacing each coordinate **r** by -**r**. If the direction of the electric field **E** is reversed in such a crystal, then the reversed field still finds an identical crystal structure and the magnitude of the induced polarisation remains unchanged;

for **E**, the polarisation is $\quad \mathbf{P}_{NL} = \varepsilon_0 \sum_m \chi^{(2m)} \mathbf{E}^{2m}$

and for -**E**, it changes to $-\mathbf{P}_{NL} = \varepsilon_0 \sum_m \chi^{(2m)} (-\mathbf{E})^{2m} = \varepsilon_0 \sum_m \chi^{(2m)} \mathbf{E}^{2m}$.

The right hand sides in the two equations are the same and hence the induced nonlinear polarisation must be zero i.e. $\chi^{(2m)} = 0$. Therefore, in order to observe even-order nonlinear effects, it is necessary to use noncentrosymmetric materials. In the beginning, familiar piezoelectric crystals were used, but recently crystals with larger nonlinear response have been developed. Besides a large value of the nonlinear susceptibility, the crystal should also have other desirable properties, e.g., a high optical-power damage threshold, transparency over a large range of frequencies etc. A list of commonly used crystals and their properties relevant to nonlinear optical applications may be found in Byer (1977). New crystalline materials are continually being added to this list - β-barium borate or BBO is an important recent addition.

Organic molecules and polymers can be synthesised to have large optical nonlinearities and good mechanical, chemical, thermal and optical stability (Garito et al., 1994). Such materials are being investigated as practical alternatives to the inorganic crystals.

EXAMPLES OF SECOND ORDER NONLINEAR EFFECTS

Historically, second order nonlinear effects have played a major role in frequency conversion processes. Standard lasers are fixed frequency devices which normally emit coherent radiation at a single frequency or in some cases at a rather limited number of discrete frequencies. Continuously tunable radiation may be obtained from dye lasers which are characterised by a broad, tens of nanometers wide, gain profile; output laser wavelengths under this profile may be tuned by placing a dispersive element within the dye laser cavity. The dye lasers operate at wavelengths ranging from about 350 nm to 1.5 μm. However, one needs to use several dyes to cover this range. Moreover, the dyes used tend to be carcinogenic, and the lasers are bulky. One of the big successes of nonlinear optics in the 1960's has been the facility to extend the range over which tunable radiation could be produced. With SHG, using a dye laser to supply the input light, it has been possible to produce continuously tunable radiation down to 200 nm. Below 200 nm, absorption in the nonlinear material adversely affects the conversion efficiency. Optical parametric oscillators have been able to provide longer wavelength radiation up to about 5 μm. New all solid-state lasing materials, for example Ti-sapphire, now eliminate the need to use dye lasers and can provide relatively compact laser systems that are continuously tunable over a very wide wavelength range (Ledingham and Singhal, 1991). Second order nonlinear effects such as the electro-optic (EO) effect have found many disparate uses such as in Q-switching, mode-locking, optical communications, optical beam deflection etc. More recently, SHG has been used in measuring time widths of femtosecond laser pulses.

In the following, we first develop the theory of SHG and discuss important practical considerations which lead to the attainment of very large conversion efficiencies. Optical parametric oscillators and the EO effect will then be discussed. Some applications will also be described in each case.

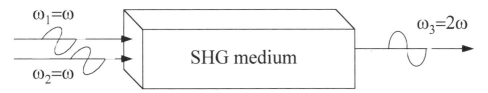

Figure 1. In second harmonic generation two photons of frequency ω are converted by the $\chi^{(2)}$ nonlinear interaction in a birefringent crystal into a photon of frequency 2ω. One method of achieving phase-matching in practice is to have the incident light polarised as an o-ray and the harmonic as an e-ray.

Second Harmonic Generation (SHG)

In SHG, two photons of frequency ω are converted by nonlinear interaction in the medium to a photon of frequency 2ω (Fig. 1). In equation 4, if we identify ω_1 and ω_2 as the initial photons of frequency ω, and ω_3 as the frequency-doubled photon of frequency 2ω, and change the other parameters for the three waves appropriately, the wave-evolution equation for 2ω may be written as

$$\frac{dE_{2\omega}(z)}{dz} = \frac{i2\omega d_{eff}\left|E_\omega(z)\right|^2}{n_{2\omega}c}e^{i\Delta kz} \ , \text{ where } \Delta k = k_{2\omega} - 2k_\omega \ . \tag{5}$$

At the input, $z = 0$, only the incident wave of frequency ω is present, i.e., $E_{2\omega}(0) = 0$. In order to study how the phase-matching condition enters into the analysis, it is convenient to study the low-conversion situation when SHG does not significantly deplete the input beam intensity and we may assume that at a distance z from the input, $E_\omega(z) = E_\omega(0)$. With these conditions, equation (5) may be integrated over values of z from 0 to z to provide the amplitude $E_{2\omega}(z)$ of the frequency-doubled wave;

$$E_{2\omega}(z) = \frac{2\omega d_{eff}\left|E_\omega(z)\right|^2}{n_{2\omega}c\Delta k}\left[e^{i\Delta kz} - 1\right] ,$$

or

$$I_{2\omega}(z) = \frac{c\varepsilon}{2n_{2\omega}}\left|E_{2\omega}(z)\right|^2 = \frac{2\omega^2\varepsilon d_{eff}^2\left(\dfrac{2n_\omega I_\omega(z)}{c\varepsilon}\right)^2}{n_{2\omega}^3c(\Delta k)^2}4\sin^2\left(\tfrac{\Delta kz}{2}\right) ,$$

or

$$I_{2\omega}(z) = \frac{8\omega^2 n_\omega^2 d_{eff}^2 z^2 I_\omega^2(z)}{n_{2\omega}^3 c^3\varepsilon}\frac{\sin^2\left(\tfrac{\Delta kz}{2}\right)}{\left(\tfrac{\Delta kz}{2}\right)^2} = \frac{8\omega^2 d_{eff}^2 z^2 I_\omega^2(z)}{n_{2\omega}^3 c^3\varepsilon_0}\frac{\sin^2\left(\tfrac{\Delta kz}{2}\right)}{\left(\tfrac{\Delta kz}{2}\right)^2} \ .$$

Hence, the conversion efficiency $\eta(z)$ for second harmonic generation is

$$\eta(z) = \frac{I_{2\omega}(z)}{I_\omega(z)} = \frac{8\omega^2 d_{eff}^2 z^2 I_\omega(z)}{n_{2\omega}^3 c^3\varepsilon_0}\frac{\sin^2\psi}{\psi^2} \ , \text{ with } \sin\psi = \frac{\Delta kz}{2}. \tag{6}$$

The conversion efficiency $\eta(z)$ varies with ψ as shown in Fig.2, being maximum for $\psi = 0$. If $\psi \neq 0$, the efficiency undergoes periodic variations as a function of ψ, the period of variation being determined by $l_{coh} = \pi/\Delta k$ which is called the phase coherence length. The nonlinear optical effects over distances of the order of an optical wavelength are usually very small, so that unless $l_{coh} \gg \lambda$, very little conversion of energy from one frequency to the other is possible. For high efficiency it is clearly important to ensure that $\psi = 0$ or, equivalently, $\Delta k = 0$. Since $\Delta k = k_{2\omega} - 2k_{\omega}$, the condition for maximum efficiency is

$$k_{2\omega} = 2k_{\omega} . \tag{7}$$

Equation (7) is an expression of the momentum conservation for the incident (also called the fundamental) and the frequency-doubled photons, and is also known as the phase-matching condition. We can rewrite this equation in terms of refractive indices for the respective waves as

$$\frac{(2\omega)n_{2\omega}}{c} = 2\frac{\omega n_{\omega}}{c} , \quad \text{or} \quad n_{2\omega} = n_{\omega} . \tag{8}$$

Equation (8) describes the index-matching condition and is an alternative statement of the phase-matching condition for second harmonic generation. It ensures that the velocities of the fundamental and second harmonic fields are equal, and there is a fixed phase relation between the induced polarisation and the generated field throughout the nonlinear medium. Under these conditions, power flows from the fundamental wave into the second harmonic. If $n_{2\omega} \neq n_{\omega}$, the direction of power flow oscillates between the fundamental wave and the SHG wave over a spatial scale characterised by $l_{coh} = \pi/\Delta k = \lambda/4(n_{2\omega} - n_{\omega})$, and no significant build-up of SHG wave intensity may occur. Therefore, for efficient SHG we require l_{coh} to be greater than the physical length l of the nonlinear medium. For $\lambda = 500$ nm and $l = 5$ cm, the refractive index mismatch must be less than 2.5×10^{-6}

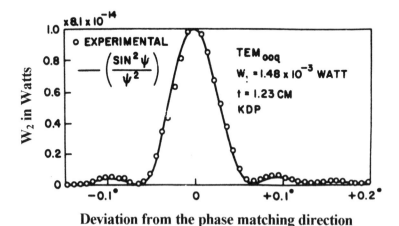

Deviation from the phase matching direction

Figure 2. Variation of SHG efficiency with the angle between the fundamental beam and the phase matching direction, as observed by Ashkin et al. (1963). The observed results agree well with the predicted dependence of $(\sin\psi/\psi)^2$.

Achievement Of Phase Matching. Away from resonances, the refractive index of a transparent medium generally decreases monotonically with increasing wavelength and may be expressed by Sellmeier's equation (where A, B, C and D are constants for the medium):

$$n^2 = A + B(\lambda^2 + C)^{-1} + D\lambda^2 . \tag{9}$$

For most materials, the refractive indices of the fundamental and the frequency-doubled waves may differ by several percent, making them unsuitable for SHG. To achieve phase matching the most popular method is to use a birefringent medium in which the refractive index depends on the polarisation state of the propagating light beam. Generally, a uniaxial birefringent crystal is used. Uniaxial crystals have a single axis of three-, four- or six- fold symmetry known as the symmetry or optic axis. The plane containing the optic axis and the direction of propagation of the light wave is called the plane of incidence. As electromagnetic waves are transverse waves, the electric vector oscillates in a plane perpendicular to the direction of propagation. The electric vector in a plane polarised light wave can be resolved into two perpendicular components - one oscillating in the plane of incidence and the other perpendicular to this plane. These components constitute two orthogonally polarised waves, respectively called the extraordinary (e-ray) and the ordinary (o-ray). The refractive indices which these polarisation waves experience are in general different. The o-ray has a refractive index which is independent of the angle θ between the direction of propagation and the optic axis. However, the e-ray experiences a refractive index which is a function of θ. In the particular case when θ = 0°, there is no distinction between the o- and the e- ray and their refractive indices are equal. For a general angle θ, the refractive index of the e-ray at frequency ω may be denoted by $n_{e,\omega}(\theta)$ and is related to

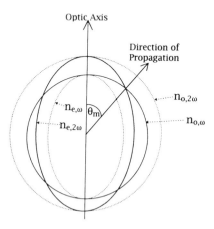

Figure 3. Practical realisation of phase matching in SHG. In the negative uniaxial birefringent nonlinear crystal, the refractive index of the fundamental beam travelling as an o-ray is equal to that of the frequency doubled beam travelling as an e-ray.

that of the o-ray according to the equation:

$$\frac{1}{n_{e,\omega}^2(\theta)} = \frac{\cos^2\theta}{n_{o,\omega}^2} + \frac{\sin^2\theta}{n_{e,\omega}^2(90^0)}. \tag{10}$$

Thus $n_{e,\omega}(\theta)$ describes an ellipse with semi-axes $n_{o,\omega}$ and $n_{e,\omega}(90^\circ) \equiv n_{e,\omega}$. The birefringence of a uniaxial crystal is defined as $n_{e,\omega} - n_{o,\omega}$, and the crystal is called negative unaxial if $n_{e,\omega} < n_{o,\omega}$ or positive uniaxial if $n_{e,\omega} > n_{o,\omega}$. The refractive indices of the o- and the e- rays change with wavelength according to equation (9), but each with their own set of constants.

Phase matching may now be achieved if the polarisation states of the fundamental and second harmonic generated waves are chosen to be orthogonal, e.g., in one scheme known as Type 1 phase matching, light of frequency ω is chosen to be the o-ray and the SHG at frequency 2ω is chosen as the e-ray. The index matching condition may then be written as $n_{e,2\omega}(\theta_m) = n_{o,\omega}$ and can be fulfilled as shown in Fig.3 for a particular choice of angle θ_m. An analytical expression for θ_m may also be obtained by rewriting equation (10) for frequency 2ω and using the phase matching condition. This yields

$$\sin^2\theta_m = \frac{n_{o,\omega}^{-2} - n_{o,2\omega}^{-2}}{n_{e,2\omega}^{-2} - n_{o,2\omega}^{-2}}. \tag{11}$$

A detailed worked example of the calculation of θ_m for the noncentrosymmetric negative uniaxial crystal, β-barium borate, has been published by Kato (1986).

The condition of the light wave travelling in the crystal at angle θ_m is very stringent and because of the dependence of $n_{e,2\omega}(\theta)$ on θ, variations of less than 0.05° may drastically reduce the conversion efficiency. The fundamental beam must be extremely well collimated, conflicting with the need to focus light to a small spot in order to achieve high intensity for increased conversion efficiencies. However, if θ_m is 90° then the variation of $n_{e,\omega}(\theta)$ with angle is much reduced (see also Fig.3) and a beam divergence of a few degrees may be tolerated. Such *non-critical phase matching* can be very useful although it is generally only possible for some discrete wavelengths. The crystal lithium niobate is used to produce frequency-doubled radiation at 532 nm from the Nd:YAG fundamental at 1064 nm using a non-critical phase matching geometry.

Wavelength-tuning in SHG. For many applications it is necessary to have a source whose wavelength is continuously tunable over a specified range. If the frequency ω is varied the phase matching condition, equation (8), will still need to be satisfied. This may be achieved either by temperature tuning or by angle tuning. Temperature tuning relies on the variation of the birefringence of the nonlinear crystal, $\Delta n(\theta,T) = n_{e,2\omega}(\theta,T) - n_{o,\omega}(T)$, on temperature T. The temperature can therefore be varied to make Δn equal to zero for the new frequency. However, angle tuning is more widely used. The crystal is rotated using step motors under computer control, while the intensity of the second harmonic generated wave is monitored by a joulemeter and the angle of the crystal set for maximum output intensity. Generally, a crystal allows only a limited range of phase matching angles to be used and a different cut of the crystal might be needed to cover a complete range of wavelengths.

In SHG the band width of the frequency doubled light is *not* increased to twice that of the incident light. This is a consequence of the fact that the conversion efficiency η depends linearly on the intensity of the fundamental beam. The spectral profile of the incident beam is either a gaussian or a lorentzian, and is characterised by a central maximum with the intensity falling sharply on either sides. The central portion of the spectral profile is, therefore, emphasised much more and light in the wings is not frequency doubled as efficiently.

The tuning range in SHG is limited to about 200 nm because of the strong optical absorption in the crystals. Below 200 nm, such absorption can cause thermal damage in the nonlinear crystal. The material β-barium borate has a high thermal damage threshold and also good transmission properties for UV radiation. In order to produce tunable radiation below 200 nm, third-order nonlinear effects are generally used. Third-order effects may be produced in a centrosymmetric medium and do not require the use of noncentrosymmetric crystals.

SHG With High Conversion Efficiency. If the conversion efficiency η is high, we can no longer ignore the depletion of the pump intensity inside the crystal. Such a situation is only likely to occur when there is proper phase matching, and for the consideration of SHG with high conversion efficiency it is reasonable to assume that $\Delta k = 0$ in equation (4). The SHG intensity produced in a nonlinear medium of length l is (Byer, 1977):

$$I_{2\omega}(l) = I_{\omega}(0)\tanh^2(\Gamma l), \quad \Gamma^2 l^2 = \frac{8\omega^2 d_{eff}^2 l^2 I_{\omega}(0)}{n_{o,\omega}^3 c^3 \varepsilon_o}. \tag{12}$$

The intensity of the fundamental beam is depleted subject to the condition $I_{2\omega}(z) + I_{\omega}(z) = I_{\omega}(0)$ at any point z in the crystal. For small values of Γ, equation (12) reduces to the phase matched case of equation (6). With an infinitely long crystal, equation (12) predicts a near 100% conversion efficiency. In practice maximum values of only up to 60 - 80% may be reached. Besides the finite dimensions of the nonlinear medium, this limitation comes primarily from the non-zero divergence of the fundamental light wave. Generally, the fundamental beam is strongly focused in the crystal in order to achieve high values of η. However, this increases the divergence of the incident beam and some parts of the beam will not be travelling in the correct direction to satisfy the phase matching condition. Typically, in a crystal of length 1 cm, a deviation of a couple of milliradians from the optimum phase matching angle is sufficient to reduce the conversion efficiency nearly to zero.

Equation (12) applies for plane incident waves. Laser beams have gaussian amplitude profile with beam divergence θ related to the minimum spot size (called the waist of the beam w_0 according to the relation, $\theta = \lambda/\pi w_0$. Moreover, there is a characteristic distance, the Rayleigh length $z_0 = \pi w_0^2/\lambda$, over which the laser spot size increases by $\sqrt{2}$. If, in order to increase the conversion efficiency, one focuses a laser beam more tightly, its divergence is increased in inverse proportion and the spot size defocuses more rapidly. This increased divergence not only causes deviations from strict phase matching but the expanding beam size results in a rapid decrease of intensity. The combination of these effects makes the conversion efficiency increase more slowly than l². An optimum choice of focusing conditions for a crystal of length l might be to have $w_0 = \sqrt{\lambda l/2\pi}$.

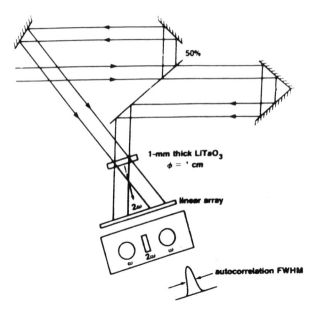

Figure 4. A single shot correlator (Maine et al., 1988) utilises SHG to measure the pulse duration of ultrashort laser pulses. The laser pulse is split into two equally intense beams which are made to overlap after travelling different path lengths. The SHG intensity is a maximum for zero path difference, but drops sharply when the beams do not overlap in the nonlinear crystal. ©1988 IEEE.

Short Pulsewidth Measurements with SHG. SHG is routinely used in laboratories to generate continuously tunable frequencies in the UV. A recent application of SHG is in measurement of the time widths of pico- and femto- second laser pulses for which electronic detectors are too slow. Here the laser pulse is split (Fig. 4) into two parts, which are

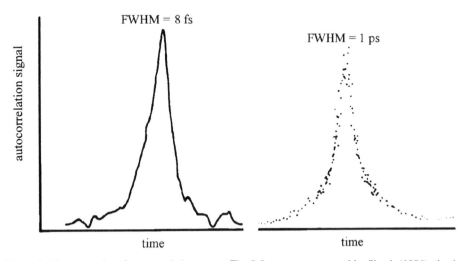

Figure 5. Two examples of autocorrelation traces. The 8 fs trace was reported by Shank (1986), the 1 ps trace (©1988 IEEE) by Maine et al. (1988).

then recombined in a nonlinear medium after they have traversed different paths. The relative delay between the two pulses may be varied. The resulting SHG pulse intensity is proportional to the product of the intensities which overlap in the nonlinear medium. For example, with no delay, there is perfect overlap and the SHG conversion efficiency has its maximum value. As the delay between the two pulses increases, the product of the intensities falls until for a delay equal to the pulse width, there is no significant overlap and SHG is much reduced. The intensity profile of the second harmonic signal is given by an autocorrelation function with the delay as the variable parameter. Two examples of such autocorrelation traces are shown in Fig.5. As the autocorrelation trace is built for each value of the delay, the measurement utilises many pulses and can only give the pulse duration for their average.

Other Three-Frequency Processes

Equations (4) describe the interaction of three plane waves propagating in a medium exhibiting second order nonlinearity. We have discussed the special case of SHG when $\omega_1 = \omega_2$ and $\omega_3 = \omega_1 + \omega_2$. Other processes which can occur with three waves are sum frequency generation (SFG), difference frequency generation (DFG) and optical parametric generation (OPG). As the names suggest, SFG produces the sum frequency $\omega_3 = \omega_1 + \omega_2$ while DFG produces the difference frequency $\omega_3 = \omega_1 - \omega_2$. In optical parametric generation, a pump wave of frequency ω_3 splits to generate two waves of lower frequencies ω_1 and ω_2. In principle, an infinite number of combinations (ω_1, ω_2) is possible. However, the energy conservation ($\omega_3 = \omega_1 + \omega_2$) and phase matching ($k_3 = k_1 + k_2$) conditions allow at most one combination of ω_1 and ω_2. If the parametric generator is placed inside a resonant cavity, it becomes an optical parametric oscillator (OPO). Such devices have found use in the production of tunable radiation over a very wide range of wavelengths. The starting point in obtaining an expression for the conversion efficiency for any of the above three-frequency processes is the consideration of equations (4). In the low conversion limit, the analysis follows similar steps to SHG. The results for SFG and DFG are as follows.

Sum Frequency Generation. When an intense laser beam of frequency ω_2 and a weak beam of frequency ω_1 are incident on a nonlinear medium, the intensity in the SFG beam at frequency $\omega_3 = \omega_1 + \omega_2$ is, in the low conversion limit,

$$I_{\omega_1+\omega_2}(z) = \frac{2(\omega_1+\omega_2)^2 d_{eff}^2 z^2 I_{\omega_2}(z) I_{\omega_1}(0)}{n_1 n_2 n_3 c^3 \varepsilon_0} \frac{\sin^2\psi}{\psi^2} \quad ; \quad \sin\psi = \frac{\Delta kz}{2} . \tag{13}$$

In the case of $\omega_1 = \omega_2$, we obtain the result of SHG. There are some interesting features in SFG now to be discussed. The phase matching condition ($\Delta k = 0$) here becomes $k_3 = k_1 + k_2$, or equivalently $n_3\lambda_3^{-1} = n_1\lambda_1^{-1} + n_2\lambda_2^{-1}$, which in combination with the energy conservation condition, $\omega_3 = \omega_1 + \omega_2$ or $\lambda_3^{-1} = \lambda_1^{-1} + \lambda_2^{-1}$, defines the index matching condition. It is often more efficient to generate tunable UV radiation by the mixing of a pump laser and the output of a dye laser than to frequency-double the dye laser output. The intensity $I_{\omega_1+\omega_2}(z)$ is proportional to the product $I_{\omega_2}(z) I_{\omega_1}(z)$ which can be up to ten times greater than is otherwise possible in SHG. The range of frequencies which may be obtained in SFG is generally wider than in SHG. Additionally, by choosing the frequencies ω_1 and ω_2 appropriately, noncritical phase matching condition ($\theta = 90°$) is easier to achieve. Besides the generation of new UV frequencies, one of the applications of SFG is in frequency up-shifting weak infrared radiation into the visible region, where detectors with much superior sensitivity are available.

Difference Frequency Generation. In DFG an intense pump wave at frequency ω_1 mixes with a second wave at frequency ω_2 to generate the difference frequency $\omega_3 = \omega_1 - \omega_2$. In the low conversion limit, the intensity in the DFG beam is

$$I_{\omega_1-\omega_2}(z) = \frac{2(\omega_1-\omega_2)^2 d_{eff}^2 z^2 I_{\omega_2}(z)I_{\omega_1}(0)}{n_1 n_2 n_3 c^3 \varepsilon_0} \frac{\sin^2\psi}{\psi^2} \quad ; \quad \sin\psi = \frac{\Delta kz}{2} . \tag{14}$$

The energy conservation condition again gives $\lambda_3^{-1} = \lambda_1^{-1} - \lambda_2^{-1}$ and the phase matching condition gives $n_3\lambda_3^{-1} = n_1\lambda_1^{-1} - n_2\lambda_2^{-1}$. A lower frequency radiation is produced, and DFG is extremely useful in producing intense tunable coherent radiation in the infrared (IR). With present nonlinear materials, it is possible to extend the tunable range of wavelengths to about 5 μm in the IR.

Optical Parametric Oscillators (OPO). In optical parametric generation an intense pump wave gives rise to a pair of lower frequency waves in a nonlinear crystal. In an OPO the nonlinear crystal is placed in an optical resonator cavity tuned to sustain either one or both of the generated waves. OPO's have been thought of as versatile sources of tunable coherent radiation capable of covering a very wide wavelength range. Commercially available OPO's can provide coverage from 400 nm to 3100 nm, and in laboratory situations they have been demonstrated to have a wavelength tuning range from 354 nm to 2370 nm (Ebrahimzadeh et al., 1990). However, the bandwidth provided by an OPO is generally much wider than that from SHG sources, and this has adversely affected the wider acceptance of OPO's. A lot of work is being done in this direction and it is hoped that with its all solid-state components and ease of operation, the OPO will in the near future be an exciting addition to the list of useful tunable sources.

If the pump wave of frequency ω_3 gives rise to a pair of signal and idler waves of frequencies ω_1 and ω_2 such that $\omega_3 = \omega_1 + \omega_2$, then in the low conversion efficiency limit for a crystal of length l, the single-pass gain at ω_1 (or ω_2) is (Harris, 1969)

$$g(l) = \gamma^2 l^2 \frac{\sinh^2\left[\left\{\gamma^2 - (\Delta k/2)^2\right\}^{1/2} 1\right]}{\left\{\gamma^2 - (\Delta k/2)^2\right\} l^2} , \tag{15}$$

where γ (the gain coefficient), ω_1 and ω_2 have been defined as

$$\gamma^2 = \frac{2\omega_0^2(1-\delta^2)d_{eff}^2 I_{\omega_3}(0)}{n_1 n_2 n_3 c^3 \varepsilon_0} , \quad \omega_1 = \omega_3(1-\delta), \omega_2 = \omega_3(1+\delta) . \tag{16}$$

The peak gain occurs at $\Delta k = 0$ and the phase matching condition may be satisfied, as in SHG, by using uniaxial birefringent noncentrosymmetric crystals with the pump wave propagating as an e-wave and the other two waves propagating as o-waves (Type I phase matching). To a good approximation, the gain falls to half of its maximum value when $|\Delta k| \sim \pi/l$. This momentum mismatch imposes a severe constraint on the pump bandwidth and in the case of a singly-resonant oscillator (SRO) resonant at the signal frequency ω_1, the maximum allowable pump bandwidth is (Young et al., 1971);

$$\Delta v_3 \left(cm^{-1} \right) = \frac{1}{l} \left[\left(n_3 - n_2 \right) + \lambda_2 \left(\frac{\partial n_2}{\partial \lambda_2} \right) - \lambda_3 \left(\frac{\partial n_3}{\partial \lambda_3} \right) \right]^{-1} . \tag{17}$$

For an XeCl excimer pump laser operating at 308 nm and a β-BaB$_2$O$_4$ crystal , equation (17) gives values in the range of 4 to 8 cm^{-1} if $\lambda_1 < \lambda_2$, and 8 to 30 cm^{-1} for $\lambda_1 > \lambda_2$. Since the spectral range of XeCl excimer radiation is several hundred wavenumbers, one has to resort to elaborate line narrowing techniques such as injection seeding. However, other lasers such as the Nd:YAG present no such problems. The constraints on the pump laser angular divergence are also quite severe and typically only a divergence of 0.2 milliradians or less is acceptable.

In order to fulfil the phase matching condition and amplify a signal wave of frequency ω_1 the nonlinear crystal must be placed such that the pump wave travels at an angle θ with respect to the crystal optic axis. This angle may be determined by consideration of the energy and momentum conservation conditions as follows:

Energy conservation $\quad\quad \omega_3 = \omega_1 + \omega_2$
Momentum conservation $\quad k_3 = k_1 + k_2$, or $n_{e,\omega_3}(\theta)\omega_3 = n_{o,\omega_1}\omega_1 + n_{o,\omega_2}\omega_2$.

Writing these equations in terms of wavelengths and reorganising, we obtain

$$n_{e,\omega_3}(\theta) = \lambda_3 \left(n_{o,\omega_1}\lambda_1^{-1} + n_{o,\omega_2}\lambda_2^{-1} \right) . \tag{18}$$

Further from equation (10), the angle θ is given as

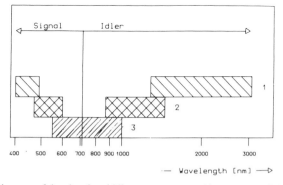

Figure 6. Wavelength range of the signal and idler waves generated in a commercial OPO system. To tune over the very wide wavelength range, three different sets of mirrors (sets 1, 2 and 3) are used.

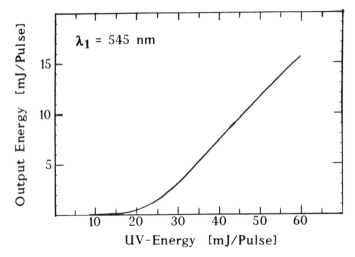

$\lambda_1 = 545$ nm

Figure 7. Spectral width of the signal wave in a commercial OPO. The spectral width depends on the pump beam energy. This dependence has been measured for a pump to threshold energy ratio equal to three.

$$\sin^2\theta = \frac{n_{e,\omega_3}^{-2}(\theta) - n_{0,\omega_3}^{-2}}{n_{e,\omega_3}^{-2} - n_{0,\omega_3}^{-2}} \ , \tag{19}$$

where the value of $n_{e,\omega_3}(\theta)$ is taken from equation (18). In order to select a different signal wavelength, it is necessary only to change the angle of the crystal. Such angle tuning provides a versatile way of scanning a large range of output wavelengths with only the crystal angle as the variable. As shown in Fig.6, the wavelength range from 400 nm to 3100 nm could be continuously tuned with a single β-BaB$_2$O$_4$ crystal and a tripled Nd:YAG laser (355 nm) as the pump laser. This figure is for a commercially available OPO whose other characteristics are shown in Figs.7 and 8.

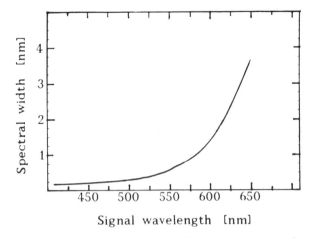

Figure 8. The energy of the signal wave as a function of pump energy at a wavelength of 355 nm.

The Linear Electo-Optic (EO) Effect

Certain materials change their optical properties when subjected to a d.c. or a low frequency a.c. electric field. By rewriting equation (1) for the instantaneous polarisation in the following form;

$$P = \mu + \left[\varepsilon_0\chi^{(1)} + \varepsilon_0\chi^{(1)}E + \varepsilon_0\chi^{(2)}E^2 + ...\right]E, \tag{20}$$

the effective polarizability of the material is seen to be $\varepsilon_0\chi^{(1)} + \varepsilon_0\chi^{(2)}E+.....$, another way of stating that the refractive index of the material is a function of E. The effective polarizability may be used to describe the effect that the material under the influence of an electric field E might exert on light passing through it. For simplicity, we assume that the external electric field is directed along the z-axis and we write the refractive index n(E) as (Shi, 1994);

$$n(E) = n_0 + a_1E + \tfrac{1}{2}a_2E^2 + ... \tag{21}$$

Generally each of the coefficients a_1, a_2 .... is a specific combination of elements of the corresponding electro-optic tensor. However, we shall not require their tensorial properties for our discussion and we shall treat them as simple variables. To the lowest order in E, equation (21) describes the linear electro-optic effect where the change in refractive index of the material is proportional to the first power of the external electric field. This is also known as the Pockels effect and the coefficient a_1 is related to the standard Pockels coefficient b_1 by the equation $b_1 = -2a_1/n_0^3$. Typical values of b_1 lie in the range 10^{-12} to 10^{-10} mV^{-1}.

The linear electro-optic coefficient is proportional to the second order nonlinear susceptibility $\chi^{(2)}$ and has a non-zero value only in noncentrosymmetric media. The EO crystal may be viewed as a capacitor across which a variable external electric field is applied. If the response of the EO crystal is to be free of distortions, the induced polarisation of the capacitor must follow the time variations in the external field. The capacitive relaxation time of the EO crystal is proportional to its dielectric constant $\varepsilon = \varepsilon_0 n_0^2$ and its value sets an upper limit on the modulation speed, which is an important parameter for information processing and transmission. A further important property of the EO material is the reduced half-wave voltage, defined as the voltage needed to be applied across a unit cube of material in order to rotate the plane of polarisation of an optical beam of wavelength λ through π radians. The reduced half-wave voltages for inorganic crystals range from about 2 to 10 kV.

The linear EO effect has found extensive applications in many disparate fields (Yariv and Yeh, 1984). In optical communications, the EO effect is used to impress information, by either amplitude- or phase- modulation, on a carrier optical beam. Commercially available EO modulators operate at speeds of 10^9 to 10^{10} Hz. Pockels cells are routinely used in Q-switching of lasers for the generation of giant pulses, and in mode-locking to generate a pulse train of short duration pulses. The linear EO effect can also be used to deflect the path of an optical beam, the extent of deflection being controlled by an external electric field.

EXAMPLES OF THIRD ORDER NONLINEAR EFFECTS

Third order nonlinear effects, mediated by the $\mathbf{E}^3$ term in equation (1), are not subject to the symmetry constraints applicable to the even order nonlinear processes, and are in principle observable in all media including liquids and gases. The polarisation induced by the $\mathbf{E}^3$ term can mix three waves in the medium and the properties of the optical wave radiated by this polarisation are described by the four-wave mixing formalism. The third order term is responsible for a rich variety of nonlinear phenomena like frequency tripling $(\omega_4 = 3\omega; \omega = \omega_{1,2,3})$, optical phase conjugation, the optical Kerr effect, self-focusing, self-phase modulation (SPM), stimulated Raman and Brillouin scattering, etc. We shall discuss some of these examples in the following.

Degenerate Four-Wave Mixing

This is a special case of the general four-wave mixing process, in which all four waves have the same frequency. If two of the waves are counter propagating ($\mathbf{k}_1 = -\mathbf{k}_2$) with the third beam incident at some arbitrary angle, then by momentum conservation (the phase matching condition), we must have $\mathbf{k}_4 = -\mathbf{k}_3$ for all directions of $\mathbf{k}_3$. We thus have a system that creates a beam ($\mathbf{k}_4$) which always retraces the path of the input beam ($\mathbf{k}_3$). This set-

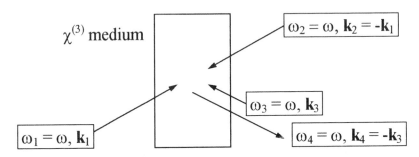

Figure 9. A schematic representation of degenerate four wave mixing.

up, Fig.9, is known as a phase conjugate mirror (PCM). The PCM returns the light wave to its source, any distortions between the source and the PCM automatically being compensated because of phase reversal. PCM's are already being used in lasers to reduce beam distortions, in adaptive optics, etc. The use of a PCM in a laser resonator cavity is demonstrated in Fig.10.

The Optical Kerr Effect

In equation (21), the term expressing the quadratic dependence of the refractive index on the electric field is responsible for the optical Kerr effect. Conventionally, for low frequency electric fields, the Kerr coefficient b_2 is defined according to the equation $b_2 = -2a_2/n_0^3$ and has values of the order of 10^{-21} m²V⁻² in liquids and 10^{-18} m²V⁻² in solids. When the electric field E is produced by an optical beam and has frequencies of the order of 10^{14} Hz, the quadratic dependence of the refractive index on E is represented as

$$\Delta_2 n = n_2 \langle E \rangle^2 = \gamma I.$$
(22)

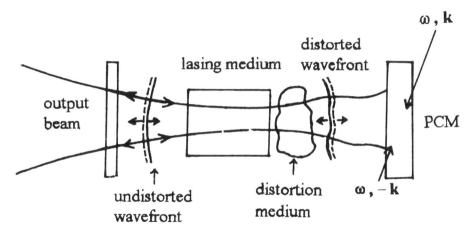

Figure 10. An arrangement showing how phase conjugation may be utilised to correct aberrations in a laser cavity. The PCM produces the phase conjugated beam which, on passing through the distorting medium, recreates the original undistorted wave front.

Here I is the light intensity and γ is called the optical Kerr effect constant. Equation (22) is the basis for controlling the propagation of an optical beam by means of a second optical beam - commonly expressed as controlling light by light or photonics. The optical Kerr effect is also responsible for laser self-focusing and self phase modulation.

Laser Self Focusing and Self Phase Modulation. Laser beams generally have a gaussian intensity profile, with the result that the light intensity is much greater on the beam axis than near the edges. According to equation (22), the refractive index and hence the effective path length of the medium will be greater along the laser beam axis. The additional phase delay along the beam axis will cause the medium to act as a converging lens to bring the light beam to a focus. For over twenty years, the phase changes due to $\Delta_2 n$ limited our ability to produce high power laser pulses and it is only in the last ten years, with the new technique of chirped-pulse amplification (CPA), that this limitation has been overcome. Essentially, the intensity within the laser amplifier must stay below a critical level, ~ 10 GW cm^{-2} both for the dyes and for solid state media. Above this level the nonlinear phase distortions in the beam produce local intensity variations which cause damage to the amplifier. For a power level of 10 GW cm^{-2}, a 1 ps long laser pulse has an energy density of only 0.01 J cm^{-2}, while solid state media, for example Nd:YAG, can support saturation energy levels of 5 J cm^{-2}. The situation gets much worse when the prevailing power levels in even shorter, sub-picosecond pulses, are considered. Therefore, the potential of solid state media as efficient laser amplifiers could not be realised for the amplification of short pulses due to the nonlinear effects, and between the years from 1968 to 1985 little progress was made in the generation of sub-picosecond pulses.

In the CPA technique, a sub-picosecond, say 100 fs, pulse is generated at the nanojoule level. The pulse is then stretched in time by a large factor of the order of 1000 or more. It is then amplified by eight to ten orders of magnitude in a solid state medium. Finally, it is recompressed to its initial pulse width. Obviously, the technology of CPA is highly sophisticated, as the stretched pulse must encode the frequency spectrum of the original pulse in the form of time within the pulse length. One of the popular methods of stretching the original sub-picosecond pulse utilises the nonlinear optical Kerr effect and is known by the name of self phase modulation (SPM).

In order to understand how SPM works, we consider a short pulse in which the intensity and hence the electric field is changing with time. Due to the nonlinear effects, the refractive index of the medium is also a function of time. We write the electric field E(t) in the light beam of central frequency ω_0 as;

$$E(t) = \Re\left[A(t)e^{i\theta(t)}\right]; \quad \theta(t) = \omega_0 t - n(t)k_0 z . \tag{23}$$

The time-dependent frequency of the wave $\omega(t)$ may then be expressed as;

$$\omega(t) = d\theta(t)/dt = \omega_0 - zk_0\,dn(t)/dt = \omega_0 - zk_0\gamma\,dI(t)/dt . \tag{24}$$

On the leading edge of the pulse, where the intensity is increasing, dI/dt is positive and the frequency is decreased below its value at the centre. On the trailing edge there is a corresponding increase in frequency, giving rise to a pulse whose frequency varies almost linearly with time over the central region of the pulse envelope. In an optical medium the refractive index n_0 changes monotonically with frequency, with different frequency components of a wave travelling at different speeds. Therefore, on leaving the nonlinear medium the frequency dispersion of the laser pulse due to SPM manifests itself as a spread in time, i.e., we obtain a chirped pulse. At the low powers of the original pulse the extent of chirping is small, but very long lengths of low loss single mode optical fibres may be used, and stretching factors of greater than 1000 are possible.

Photonics. Equation (22) is the basis on which a major part of the new technology of photonics is founded. 'Photonics' refers to the concept of computing and data transmission using photons in place of electrons. All-optical devices have many important advantages over their conventional electronic counterparts. For example, the operational speed of optical devices can be several orders of magnitude higher, as they are not limited by circuit inductances and capacitances. Electronics communication devices typically have bandwidths of less than 1 GHz, while optical devices are characterised by bandwidths in the region of 10^{14} Hz. Separate light beams can occupy the same space in a linear optical medium while electronic circuits need separate wires for connections. Conservatively, up to 10^4 more connections may be made in the same space using light beams than with electronic devices. Moreover, in optical devices heat dissipation is not a problem. At present, the nonlinear

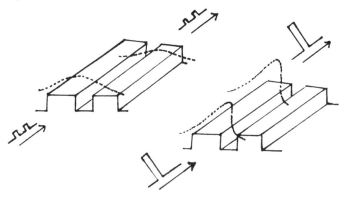

Figure 11. A nonlinear directional coupler consists of two waveguides sufficiently close for the evanescent field of an optical pulse to overlap across them. A low intensity pulse in one waveguide is coupled across to the neighbouring one. However, a high intensity pulse changes the refractive index of the medium and the coupling to the neighbouring waveguide is blocked with the result that the high intensity pulse is transmitted in the same waveguide.

effects in commonly used materials are weak, requiring large switching energies. Considerable progress is being made in synthesising materials based on organic and polymer systems which may have one to two orders of magnitude larger optical nonlinearities.

Fig.11 shows a nonlinear directional coupler consisting of two waveguides that are physically so close to each other that the evanescent field of an optical pulse will overlap the two waveguides, a low intensity light pulse in one waveguide being coupled across to the other. However, a high intensity pulse causes a change in the refractive index of the medium with the result that such a pulse is confined to the same waveguide through total internal reflections. This arrangement can be used to sort a series of weak and strong pulses and separate them into the two outputs.

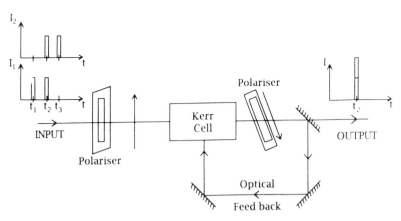

Figure 12. An optical logic AND gate.

As a final example of application of the optical Kerr effect, we describe the principle of operation of an optical AND logic gate. If polarised light passes through a Kerr medium, then by using control light to change the refractive index of the medium, it is possible to rotate the plane of polarisation of the light beam. Therefore, under the effect of the control light, such a Kerr cell placed between a pair of crossed polarisers will control the light intensity transmitted by the second polariser. The control light could be derived by partially reflecting the original light pulse at the output of the Kerr cell and feeding the reflected light back into the cell (Fig.12). When the incident light intensity is low, the nonlinear modification of the refractive index is small and consequently the transmitted intensity is blocked by the crossed polariser. However as the intensity of the incident light increases, the efficiency of optical feedback also undergoes an increase, causing a bigger change in the refractive index. This in turn rotates the plane of polarisation of the output pulse by progressively larger amounts: more light passes through the second polariser, and as a result a larger intensity is fed back into the cell. This sets up a cascade, and the output jumps to a level for maximum transmission through the polariser. At the end of the input pulse, the light falls back to a low level and the polariser cuts out the transmitted light. The onset of transmission and extinction of light through the polariser takes place at specific light intensity values, with the whole cycle forming a hysteresis loop. If the individual pulse intensities are chosen such that on their own they do not cause transmission, but together an output pulse does result, then the device as described above acts as an optical logic AND gate.

HIGHER ORDER NONLINEAR EFFECTS

In Equation (1), we had expressed the induced polarisation **P** as a power series in **E**, the external applied electric field. One can show in general that the ratio of the successive terms in this expansion are typically (Shen, 1984);

$$\left| \frac{P^{(n+1)}}{P^{(n)}} \right| = \left| \frac{\varepsilon_0 \chi^{(n+1)}}{\varepsilon_0 \chi^{(n)}} \right| = \left| \frac{E}{3 \times 10^{10} \, Vm^{-1}} \right|. \tag{25}$$

For conventional nanosecond lasers, the electric field strengths are of the order of 10^8 Vm^{-1} and the ratio of successive terms is ~0.01. This makes the series convergent. With the availability of femtosecond lasers (Perry and Mourou, 1994), focused intensities of 10^{20} W m^{-2} are routinely realised. These correspond to electric fields of 3×10^{11} Vm^{-1}, and the ratio of successive terms in equation (1) is 10. The perturbation expansion of (1) is clearly not valid for such high laser intensities and higher order nonlinear polarisations could be important. That this indeed is the case is demonstrated by the observation of extremely high odd-order harmonics with intense lasers. Figure 13 shows the harmonic spectrum (Macklin et al., 1993) obtained with 125 fs laser pulses of 800 nm wavelength at focused intensities of 10^{19} W m^{-2}. The target was neon, with an effective thickness of 32 mm-torr. It is seen that harmonic generation of order up to 111 takes place, with peak intensity occurring for harmonic orders in the fifties. Additionally it should be noted that radiation of wavelengths as low as 7.5 nm is being produced from an 800 nm incident wavelength! The coherence and overall conversion efficiency of the extreme ultraviolet (XUV) and soft X-ray radiation are limited by phase matching between the harmonic field and the incident laser field. If the Rayleigh length of the focused beam is much longer than the length of the nonlinear medium, then low-divergence harmonic beams with a near-gaussian spatial distribution can be produced, and conversion efficiencies of 10^{-7} to 10^{-6} achieved.

The relation expressed by equation (25) is valid only in situations where the incident laser frequency is away from atomic/molecular resonances. Near resonances, energy difference denominators in the nonlinear susceptibilities may greatly enhance the importance

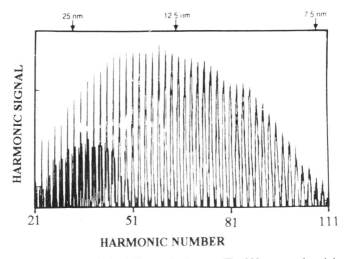

Figure 13. The production of very high odd harmonics in neon. The 800 nm wavelength laser pulse has an intensity of 10^{19} W m^{-2}.

of higher order terms, and nonlinear effects may be observable with much weaker optical beams. Non-resonant multiphoton ionisation of atoms and molecules can happen with intense laser beams, but if the laser frequency is in resonance with one of the transitions in the medium, the probability of excitation/ionisation is increased by orders of magnitude. Nonlinear laser spectroscopy is a vast field with many practical applications, and there is not scope to discuss this branch of nonlinear optics further in this introductory article.

CONCLUSIONS AND FUTURE OUTLOOK

Nonlinear optical effects, in the form of the Pockels and Kerr effects, were observed with static or low frequency fields about a hundred years ago, but the real advances have been made since the advent of lasers in 1960. With the availability of very high power femtosecond lasers, the science of nonlinear optics is set to expand as new, sometimes bizarre effects, manifest themselves at such incredibly high electric field strengths. For example, at light intensities of 10^{25} W m^{-2} the prevailing electric field is sufficient to strip up to 82 electrons from a uranium atom, and the radiation pressure is 300 Gbar!

The new technology of photonics aims to develop all-optical switches and all-optical logic elements. Even now, it is possible to place an array of 100×100 pixels on a 1 cm^2 GaAs chip (Shi, 1994). If the switching time is set at 10^{-10} s then an operation rate of 10^{14} bit s^{-1} can be realised, which is four orders of magnitude faster than the rate at which electronic supercomputers can operate. The major problem at this stage of development is the requirement of the large switching energy of about 65 picojoules needed to activate the optical nonlinearity. This switching energy must be lowered to a value of less than a picojoule per bit for the all optical device to be commercially attractive. A lot of effort is being put into the development of novel organic and polymeric materials which exhibit large optical nonlinearities.

The research field of nonlinear optics is expanding very rapidly, with numerous applications in disparate fields. The availability of compact optical sources and materials with large optical nonlinearities will bring about the introduction of faster, smaller and versatile devices.

REFERENCES

Ashkin, A., Boyd, G. D., and Dziedzic, J. M., 1963, *Phys. Rev. Letts* 11:14.
Byer, R.L., 1977, *in*: "Proceedings of 16th Scottish Universities Summer School In Physics 1975," P.G. Harper and B.S. Wherrett, eds., Academic Press, ew York.
Ebrahimzadeh, M., Henderson, A. J., and Dunn, M. H., 1990, *IEEE J. Q. Elec.* QE-26:1241.
Garito, A., Shi, R.F., and Wu, M., 1994, *Physics Today* 47(5):51.
Harris, S. E., 1969, *in*: "Tunable Optical Parametric Oscillators," Proc. IEEE 57:2096.
Kato, K., 1986, *IEEE J. Q. Elec.* QE-22:1013.
Ledingham, K.W.D., and Singhal, R.P., 1991, *J. Anal. Atomic Spectroscopy* 6:73.
Macklin, J. J., Kmetec, J. D., and Gordon III, C. L., 1993, *Phys. Rev Letts* 70:766.
Maine, P., Strickland, D., Bado, P., Pessot, M., and Mourou, G., 1988, *IEEE J. Q. Elec.* QE-24:398.
Perry, M. D., and Mourou, G., 1994, *Science* 264:917.
Shank, C.V., 1986, *Science* 233:1276.
Shen, Y. R. 1984, "The Principles Of Nonlinear Optics," Wiley, New York.
Shi, Shu, 1994, *Contemp. Phys.* 35:21.
Yariv, A., 1985, "Optical Electronics," Holt-Saunders, Japan.
Yariv, A., and Yeh, P., 1984, "Optical Waves in Crystals," Wiley, New York.
Young, J. F., Miles, R. B., Harris, S. E., and Wallace, R. W., 1971, *J. Appl. Phys.* 42:497.

APPLICATIONS OF SURFACE SECOND ORDER NONLINEAR OPTICAL SIGNALS

Michael J. E. Morgenthaler and Stephen R. Meech

School of Chemical Sciences
University of East Anglia
Norwich NR4 7TJ

INTRODUCTION

In this chapter the theory and practice of the application of second order nonlinear optical signals to surface spectroscopy and kinetics will be described. The vast majority of nonlinear optical spectroscopic methods which were developed when high peak power pulsed lasers became available exploit the third order nonlinear susceptibility, $\chi^{(3)}$. Second order nonlinear optical signals are largely of interest as a tool for frequency shifting laser output in anisotropic nonlinear crystals such as KDP (potassium dihydrogen phosphate) by second harmonic generation, optical parametric conversion, etc. The spectroscopic application of second order signals has been neglected for the very good reason that $\chi^{(2)}$ is zero by symmetry in isotropic and other centrosymmetric media, which are the media of most interest in spectroscopy. However, there is one area in which this symmetry selection rule is a positive advantage, namely surface spectroscopy.

At the interface between two media, inversion symmetry is necessarily absent. This means that at the interface second order signals, for example second harmonic generation (SHG), are symmetry allowed. If the two bulk media on either side of the interface are themselves isotropic then any second harmonic signal observed arises exclusively from the interface. This surface SHG (or sum frequency generation, SFG) represents an *all optical surface-specific* signal. The all optical (photon in - photon out) nature of the signal gives it a considerable advantage over other surface-specific spectroscopies, such as electron diffraction or electron energy loss, which require one of the media to be a vacuum. In contrast SHG can be measured even at liquid/liquid interfaces. The surface-specific nature of the signal is a great advantage over conventional, linear, spectroscopic measurements which can be applied to interfaces only when the surface excess of the adsorbate is very large. In contrast SHG and SFG can distinguish surface and bulk species even when they comprise the same molecule.

These unique attributes of surface second order nonlinear optical signals have attracted attention from a wide variety of chemists and physicists. Electrochemists have used the SHG intensity as a function of applied potential to monitor a variety of electrode processes (Corn

An Introduction to Laser Spectroscopy
Edited by David L. Andrews and Andrey A. Demidov, Plenum Press, New York, 1995

and Higgins, 1994; Richmond et al., 1988). Surface scientists have been able to determine the structure of clean surfaces using polarisation resolved measurements (Heinz et al., 1985), while surface chemists have monitored SHG intensity as a function of exposure to measure adsorption isotherms (Shen, 1986). Spectroscopists have employed infra-red and visible sum frequency generation to record the vibrational spectroscopy of adsorbates (Harris et al., 1989). Ultrafast spectroscopists have used surface SHG as a probe in picosecond pump and probe measurements to monitor the fast reactions of adsorbates (Sitzmann and Eisenthal, 1988; Meech and Yoshihara, 1989).

In the following section the theory of surface second harmonic generation will be described, leading on to a discussion of how the technique can be used to measure adsorbate spectra and adsorbate orientation. In the third section some of the experimental techniques mentioned above will be described in more detail. The main purpose of the following sections is to provide an introduction to surface nonlinear optical spectroscopy; readers interested in a more detailed picture can consult several more comprehensive reviews, (Heinz, 1991; Eisenthal, 1993; Andrews, 1993; Corn and Higgins, 1994; Meech, 1993).

SURFACE NONLINEAR OPTICAL SIGNALS

In the following we will briefly outline the derivation of a general expression for the second harmonic (or sum frequency) signal generated by an interface with a second order nonlinear susceptibility tensor $\chi_s^{(2)}$, illuminated by radiation at frequency ω (or ω_1 and ω_2 for SFG). This result will then be studied for its behaviour as a function of input and output polarizations. It will be shown how, for certain surface structures, individual elements of the second rank tensor $\chi_s^{(2)}$ can be determined: the experimental apparatus is shown in figure 1.

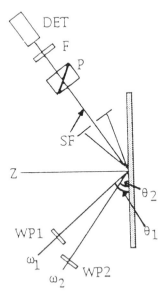

Figure 1. General experimental arrangement for surface second order nonlinear optical measurements. Laser pulses of controllable polarisation at ω_1 and ω_2 are overlapped on the surface in space and time. The sum frequency signal is reflected in the phase matched direction, spatially and spectrally filtered from the fundamental beams and observed by the detector through an analysing polariser. In the case of SHG measurements only a single incident laser pulse is required.

These elements are related to elements of the microscopic quantity, the molecular hyperpolarisabilty tensor, β. These relationships may be exploited to determine the mean orientation of the adsorbate. A consideration of the resonance behaviour of β also reveals the origin of adsorbate spectra in IR-plus-visible sum frequency generation.

The Signal Intensity

The macroscopic picture of optical signals is an applied electric field polarising a dielectric medium which itself radiates. If the electric field is very strong, nonlinear terms in the polarisation become important.

$$\mathbf{P}^{(1)}(\omega) = \chi^{(1)}\mathbf{E}(\omega) , \tag{1a}$$

$$\mathbf{P}^{(2)}(2\omega) = \chi^{(2)}\mathbf{E}(\omega)\mathbf{E}(\omega) , \tag{1b}$$

$$\mathbf{P}^{(3)}(3\omega) = \chi^{(3)}\mathbf{E}(\omega)\mathbf{E}(\omega)\mathbf{E}(\omega) . \tag{1c}$$

The polarisation of interest here is $\mathbf{P}^{(2)}(2\omega)$ in (1b). The origin of the polarisation, and hence radiation, at 2ω is clear if we write $E(\omega) = E_0\cos(\omega t)$, in which case 1b may be written in scalar form as:

$$P^{(2)}(2\omega) = \chi^{(2)}\frac{E_0^2}{2}\left[1 + \cos(2\omega t)\right] . \tag{2}$$

It is also apparent from (1b) that $\chi^{(2)}$ must be zero in an isotropic medium. If the direction of the field is reversed, then by symmetry the direction of polarisation must also reverse. However, since $(-E(\omega))^2 = E^2(\omega)$ this does not happen. The only acceptable solution is $\chi^{(2)} = 0$ in isotropic media.

To treat specifically the case of an interface we may begin with the expression for the field radiated by a polarised slab of thickness smaller than the wavelength of incident radiation (Heinz, 1991):

$$\hat{\mathbf{e}}_i \cdot \mathbf{E}_i = \frac{2\pi i\Omega}{\varepsilon_i^{1/2}c}\sec\theta_i\left(\mathbf{e}\cdot\mathbf{P}_s\right) , \tag{3}$$

where θ_i is the angle between the surface normal and a beam radiated at frequency Ω, and $\hat{\mathbf{e}}_i$ is a polarisation vector of the radiated field in medium i. The polarisation $\mathbf{P}_s$ has the form:

$$\mathbf{P}_s(\Omega) = \chi_s^{(2)}(\Omega){:}\mathbf{e}_i(\omega_1)\mathbf{e}_i(\omega_2)E_i(\omega_1)E_i(\omega_2) , \tag{4}$$

where E_i are the incident field amplitudes, $\Omega = \omega_1 + \omega_2$, and the $\mathbf{e}_i$ are products of the pump polarisation vectors and their corresponding Fresnel coefficients for transmission at a dielectric boundary (Jenkins and White, 1981). By combining equations (3) and (4) and expressing the result in terms of the power at Ω the following important result emerges:

$$I(\omega) = \frac{8\pi^3\Omega^2\sec^2\theta}{c^3\left[\varepsilon_i(\Omega)\varepsilon_i(\omega_1)\varepsilon_i(\omega_2)\right]^{1/2}}\left|\mathbf{e}(\Omega)\cdot\chi_s^{(2)}{:}\mathbf{e}(\omega_1)\mathbf{e}(\omega_2)\right|^2 I(\omega_1)I(\omega_2) . \tag{5}$$

Equation (5) is the basic result of the theory of surface second order nonlinear optical phenomena. It was originally derived by Heinz (1982). Mizrahi and Sipe (1988) used a slightly different model to obtain a very similar result, and showed that the result is valid for a variety of geometries, provided the Fresnel factors are properly evaluated. The result is valid for both real and complex dielectric constants. Equation (5) shows that $I(\Omega)$ depends linearly on the intensity of each input beam (or quadratically on intensity for SHG) and quadratically on the magnitude of the nonlinear susceptibility. Further, the intensity will clearly be a strong function of the geometry of the experiment and the polarisation of the input beams, via the tensor character of $\chi_s^{(2)}$, the Fresnel factors and θ_i. The direction of the reflected (or transmitted) output beam is determined by the input geometry. Coplanar, non-collinear pump beams, incident at angles θ_1 and θ_2, give an SFG output at $\Omega_i(\Omega)$ where:

$$n(\Omega_i)\Omega \sin \theta_i = n(\omega_1)\omega_1 \sin \theta_1 + n(\omega_2)\omega_2 \sin \theta_2 \ . \tag{6}$$

For SHG, since $\omega_1 = \omega_2$, $\theta_1 = \theta_2$, the reflected signal is collinear with the reflected fundamental provided the medium is dispersionless. Where the medium through which the signal is observed exhibits dispersion, the fundamental and second harmonic beams are spatially separated.

It is straightforward to express equation (5) in terms of the number of photons at the second harmonic frequency, $S(2\omega)$, if the surface nonlinear susceptibility is represented by some effective value $\chi_{s,eff}^{(2)}$;

$$S(2\omega) = \left(\frac{16\pi^3 \omega}{\hbar c^3}\right) \frac{P^2}{A\tau R} \left|\chi_{s,eff}^{(2)}\right|^2 \ . \tag{7}$$

The incident beam at frequency ω is characterised by a pulsewidth τ, repetition rate R, and power P; the area illuminated is denoted by A. An approximate calculation reveals the magnitude of $\chi_{s,eff}^{(2)}$ that can yield an observable signal. For example a 10 ns pulsewidth 10 Hz Nd:YAG laser with 1 W output power, illuminating a 1 cm^2 area of a surface with $\chi_{s,eff}^{(2)}$ equal to 10^{-17} esu, will yield 10 photons per second at 532 nm. This certainly represents a small signal, but all of the photons in the collimated signal beam can be collected, and gated detection can be used to reduce background. A $\chi_{s,eff}^{(2)}$ value of 10^{-17} esu would be appropriate for a monolayer of weakly nonlinear molecules in the absence of resonance enhancement. Resonant adsorbates (i.e. those with an electronic transition near ω or 2ω) on insulator substrates and clean semiconductor or metal surfaces usually exhibit a much larger effective nonlinear susceptibility, so the signal may readily be detected.

From equation (7), it follows that $S(2\omega)$ is maximised when all of the available energy is concentrated in a tightly focused ultrashort pulse. However such experiments are limited by the damage threshold of the surface. In addition many of the 'quasi cw' time-resolved experiments described below are not feasible with a low repetition rate giant pulse laser. In such cases a quasi-cw mode-locked Nd:YAG laser (P = 10 W, τ = 100 ps R = 100 MHz) is popular. For a 10 photon s^{-1} signal from the same surface the cw laser must be focused to a 10^{-3} cm^2 spot size. However a tenfold reduction in peak power is achieved, reducing the risk of surface damage. For either laser it should be noted that surface heating (either transient or continuous) can present a problem. More recently quasi-cw femtosecond lasers, such as Ti:Al$_2$O$_3$, have been used, providing larger signals for an even lower average power.

Determining the Elements of $\chi_s^{(2)}$

From figure 1 it is easy to see that a large number of distinct measurements of $I(2\omega)$ can be made. Most straightforwardly the analysing polariser can select either the p or s polarised SHG signal. Any combination of input polarisations can be obtained by varying the rotation angle of the $\lambda/2$ plate. For simplicity only the (single input beam) case of SHG will be considered here, in which case the $\lambda/2$ plate varies the polarisation angle α. In addition, though these parameters are less easy to vary experimentally, the angle of incidence and angle of rotation of the surface about its normal may also be changed. It is easier to see how such variations influence $I(2\omega)$ by expanding out the tensor $(\chi_s^{(2)})$ and vector $(\mathbf{E}, \mathbf{e})$ quantities of equations (4) and(5) into their individual elements. The final objective is to determine individually the elements of $\chi_s^{(2)}$, since these lead to microscopic information about the adsorbate.

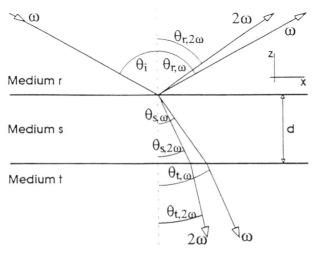

Figure 2. Three layer model of an interface. The angles indicated are required, along with the polarisation angle, in calculation of the fields in the interface (s). For further details see text.

The starting point for these calculations is the three layer geometry of figure 2, and the equations originally derived by Bloembergen and Pershan (1962) for the p and s polarised electric fields radiated by a polarised slab; these have the general form of equation (2) but explicitly require the P_i, where i = x, y, z are the Cartesian coordinates of the surface (z is the normal, zx the plane of incidence). The task is to calculate P_i, for which we require knowledge of both $\chi_s^{(2)}$ and the electric field in the interface.

The second rank tensor $\chi_s^{(2)}$ comprises 27 elements $\chi_{ijk}^{(2)}$. Clearly for most applications the associated complexity has to be reduced. For SHG the index symmetry $\chi_{ijk}^{(2)} = \chi_{ikj}^{(2)}$ may be used. If the symmetry of the interface is known, or may be assumed, many of the remaining 18 elements may reduce to zero, or turn out to be equal to other elements. The non-zero elements for several different interface symmetries have been tabulated by Heinz (1991). A particularly common symmetry is one with 2D isotropy in the surface plane. This is expected for adsorbates at liquid/vapour and non-crystalline solid/isotropic fluid interfaces. Such symmetry can be asssumed if $I(2\omega)$ is observed to be invariant to rotation about the surface normal. Application of the relevant symmetry operations to $\chi_s^{(2)}$ leaves only three non-zero independent elements, and the polarisations P_i are easily calculated:

$$
\begin{vmatrix} P_x \\ P_y \\ P_z \end{vmatrix} = \begin{vmatrix} 0 & 0 & 0 & 0 & \chi_{xzx} & 0 \\ 0 & 0 & 0 & \chi_{xzx} & 0 & 0 \\ \chi_{zxx} & \chi_{zxx} & \chi_{zzz} & 0 & 0 & 0 \end{vmatrix} \begin{vmatrix} E_x^2 \\ E_y^2 \\ E_z^2 \\ 2E_zE_y \\ 2E_zE_x \\ 2E_xE_y \end{vmatrix} . \tag{8}
$$

Equation (8) makes possible a number of useful predictions. If the fundamental beam is s-polarised, $P_x = P_y = 0$ and $P_z = \chi_{zxx}E_y^2$ so that $I(2\omega)$ is p-polarised and dependent on a single tensor component. Similarly a p-polarised fundamental beam leads to a purely p-polarised $I(2\omega)$. An s-polarised signal is only detected with a mixed input polarisation. Further, all non-zero elements of $\chi_s^{(2)}$ contain a z index, meaning that for normal incidence $I(2\omega) = 0$. Such polarisation selection rules are useful in characterising the SHG signal.

In equation (8) the fields E_i are the fields in the interfacial layer s of figure 2. These must be calculated as a function of polarisation angle;

$$
\begin{vmatrix} E_x \\ E_y \\ E_z \end{vmatrix} = \begin{vmatrix} a\cos\alpha \\ c\sin\alpha \\ b\cos\alpha \end{vmatrix} \mathbf{E_0} , \tag{9}
$$

where the factors a, b and c are the effective Fresnel coefficients for transmission of the fundamental beam from medium r to s. These have been tabulated for the geometry of figure 2 (Shen, 1989) for the simpler single interface system (Mizrahi and Sipe, 1988) and more complex interfaces (Grudzkov and Parmon, 1993). Equations (8) and (9) may be combined to obtain P_i which, on substitution into the equations of Bloembergen and Peshan (1962) (see also Zhang et al., 1990), yield results of the following form for the s- and p- polarised $I(2\omega)$:

$$
I_p(2\omega) = \left(A\cos^2\alpha + B\sin^2\alpha \right)^2 I^2(\omega) , \tag{10}
$$

$$
I_s(2\omega) = \left(C\sin^2\alpha \right)^2 I^2(\omega) , \tag{11}
$$

$$
A = a_1\chi_{xzx} + a_2\chi_{zxx} + a_3\chi_{zzz} , \tag{12}
$$

$$
B = a_4\chi_{zxx} , \tag{13}
$$

$$
C = a_5\chi_{xzx} , \tag{14}
$$

where the a_i of equations (12) - (14) are functions of the angles and refractive indices of the media represented in figure 2. By evaulating these a_i for a given θ_i, and curve fitting to (10) and (11) to obtain results from (12) - (14), the three independent elements of $\chi_s^{(2)}$ are obtained. An example of such a fit is shown in figure 3. The major problem with this type of measurement lies in the value assumed for the dielectric constant of the interfacial region, which in the (usual) case of a resonantly enhanced signal will be complex. Different model assumptions for n_s have been discussed by Grudzkov and Parmon (1993): nonetheless it should be borne in mind that this unknown factor limits the reliability of $\chi_s^{(2)}$ determinations.

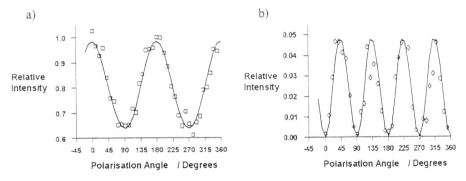

a)

b)

Figure 3. SHG intensity, (a) p-polarised and (b) s-polarised, measured as a function of polarisation angle for a monolayer of rhodamine 110 on silica. Solid lines are fits to the data points using equations (10) and (11). The fundamental wavelength was 532 nm.

Applications of SHG in Determining Adsorbate Orientations, Kinetics and Spectra

The key relationship between the measured, macroscopic $\chi_{ijk}^{(2)}$ and the microscopic, molecular, parameters of interest is

$$\chi_{ijk}^{(2)} = N_s \langle T_{ii'} T_{jj'} T_{kk'} \rangle \beta_{i'j'k'} \ , \tag{15}$$

where the β_{ijk} are the 27 elements of the molecular hyperpolarisability tensor. The prime indicates the molecular frame, the z' axis being taken as the principal symmetry axis of the adsorbate. The matrix $(T_{\lambda\lambda'})$ transforms between the surface and molecular frames, and the number density of adsorbates is N_s.

Equation (15) contains most of the information about an adsorbed monolayer that may be extracted from measurements of I(2ω): adsorbate orientation, adsorbate kinetics and adsorbate spectroscopy. These will be considered in turn.

Orientation. The transformation between laboratory and molecular frame can be achieved by describing the relative orientation of the frames in terms of Euler angles. The principle is illustrated in figure 4; θ is the angle between the surface normal and z', ψ the angle through which the molecule is twisted about its z' axis and φ the angle between the y axis and the projection of z' onto the surface. The product of the three 3×3 matrices required to obtain alignment of each axis yields the required expression for $(T_{\lambda\lambda'})$ (Goldstein, 1969). This can then be applied to the tensor β. The number of non-zero elements of β is reduced by the symmetry of the adsorbate, in much the same way as for $\chi_s^{(2)}$. For C_{2v} symmetry there are 7 non-zero elements. For resonant SHG (see below) with planar aromatic adsorbates the dominant terms in the hyperpolarisability are in plane, so elements containing the y' index can be set to zero. Three independent non-zero elements remain: $\beta_{z'z'z'}$, $\beta_{z'x'x'}$ and $\beta_{x'x'z'}$. Multiplying the matrices $(T_{\lambda\lambda'})$ with β_{ijk} leads to an expression for each non-zero $\chi^{(2)}$. As the resulting equations contain too many parameters to allow adsorbate orientation to be determined, they must be simplified further. If the second harmonic frequency is resonantly enhanced by an electronic transition of known polarisation it is possible to reduce the number of significant elements of β to one or two (Corn and Higgins, 1994). It is also necessary to make some model assumptions about the angle ψ whenever significant elements contain the x' index. Sensible models are; (a) a random orientation or (b) that the y' axis lies parallel to

177

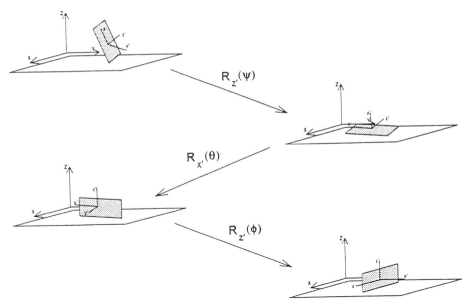

Figure 4. Sequential operations required to transform the molecular to the laboratory frame: the **R** matrices are given in Goldstein, 1969.

the surface ($\psi = 90°$). Whether (a) or (b) is valid depends on the molecular structure of the adsorbate. With these simplifications it is possible to obtain a simple relationship between the measured $\chi_{ijk}^{(2)}$ and the molecular tilt angle θ. Examples are shown in table 1.

It should always be remembered that ultimately the usefulness of the orientation parameter $D = <\cos^3\theta>/<\cos\theta>$ (table 1) depends on the validity of the following assumptions: (1) azimuthal symmetry at the surface, reducing the number of significant components $\chi_{ijk}^{(2)}$, (2) correct estimation of the interfacial dielectric constant ε_s, (3) adsorbate symmetry the same as that of the free molecules, (4) resonance enhancement ensuring that only 1 or 2 elements of β are significant, and (5) the assumption that the model of ψ is reasonable. The likelihood that all of these assumptions are rigorously obeyed is small for all but the simplest systems; measurements of D are therefore more useful in a relative (e.g. as a function of coverage) than an absolute sense. A final assumption, namely that $D = \cos^2\theta$, requires a very sharply peaked orientational distribution. This will only be true for very well defined surfaces.

Kinetics. Essentially adsorbate kinetics are measured by recording the square root of the nonlinear signal intensity as a function of time; from equation (15), $[I(2\omega)(t)]^{1/2} \propto \chi_{s,eff}^{(2)}(t) \propto N_s(t)$. The relationship is obviously more complicated if adsorbate orientation changes along with N_s, but from the preceding discussion it is clear that this possibility can be eliminated by measuring the kinetics for different $\chi_{ijk}^{(2)}$. A complication that is not so easily neglected is two distinct (e.g. adsorbate and substrate) sources of SHG. Since the two sources need not have the same phase, the possibility of interference between them must be considered. In general,

$$\chi_{s,eff}^{(2)} = A(1-\Theta) + B\Theta + C\Theta , \tag{16}$$

where A is the nonlinear susceptibility of the bare surface, B that of the adsorbate covered surface, C that of the adsorbate itself and Θ the fractional coverage, N_s/N_∞. For

Table 1. The three nonzero elements of $\chi_s^{(2)}$ may be combined to obtain the orientation parameter D. The appropriate combination is a function of the number of non-zero elements of the molecular hyperpolarisability and the model assumed for the angle ψ.

Dominant terms of $\beta^{(2)}$	ψ	D
β_{zzz} only	Random or 90°	$\dfrac{\chi_{s,zzz}^{(2)}}{2\chi_{s,zxx}^{(2)} + \chi_{s,zzz}^{(2)}}$
β_{zxx} only	Random	$\dfrac{2\chi_{s,zxx}^{(2)} - \chi_{s,zzz}^{(2)}}{2\chi_{s,zxx}^{(2)} + \chi_{s,zzz}^{(2)}}$
	90°	$\dfrac{2\chi_{s,zxx}^{(2)}}{2\chi_{s,zxx}^{(2)} + \chi_{s,zzz}^{(2)}}$
β_{xzx} only	Random	$\dfrac{\chi_{s,xxz}^{(2)}}{\chi_{s,xxz}^{(2)} - \chi_{s,zxx}^{(2)}}$
	90°	$\dfrac{2\chi_{s,xxz}^{(2)} - \chi_{s,zxx}^{(2)}}{2\chi_{s,xxz}^{(2)} - 2\chi_{s,zxx}^{(2)}}$
β_{zzz} and β_{zxx}	Random	$\dfrac{\chi_{s,zzz}^{(2)} - \chi_{s,zxx}^{(2)} + \chi_{s,xxz}^{(2)}}{\chi_{s,zzz}^{(2)} - \chi_{s,zxx}^{(2)} + 3\chi_{s,xxz}^{(2)}}$
	90°	$\dfrac{\chi_{s,zzz}^{(2)} - 2\chi_{s,zxx}^{(2)} + 2\chi_{s,xxz}^{(2)}}{\chi_{s,zzz}^{(2)} - 2\chi_{s,zxx}^{(2)} + 4\chi_{s,xxz}^{(2)}}$
β_{zxx} and β_{xzx}	Random	$\dfrac{2\chi_{s,zxx}^{(2)} + \chi_{s,zzz}^{(2)} + 4\chi_{s,xxz}^{(2)}}{2\chi_{s,zxx}^{(2)} + 3\chi_{s,zzz}^{(2)} + 4\chi_{s,xxz}^{(2)}}$
	90°	$\dfrac{2\chi_{s,zxx}^{(2)} + 2\chi_{s,zzz}^{(2)} + 4\chi_{s,xxz}^{(2)}}{2\chi_{s,zxx}^{(2)} + 3\chi_{s,zzz}^{(2)} + 4\chi_{s,xxz}^{(2)}}$

the completely general case of a significant contribution from A, B and C, each with a different phase, the analysis is quite complicated. Fortunately, it is often possible to neglect one or more of the contributions to (16). For example, when C is finite and A = B = 0, which may occur for an aligned resonant dipolar adsorbate on a weakly nonlinear substrate, equation (16) yields $E(2\omega) \propto \chi_{s,eff}^{(2)} \propto C\Theta$, where the constant $C = N_\infty \langle \beta_{s,eff} \rangle$. The signal intensity then directly yields the adsorbate coverage. From time-dependent measurements, adsorption isotherms, or even picosecond adsorbate kinetics, are obtained.

Another widely studied case is that of a weakly nonlinear adsorbate on a metal surface. The largest contribution to the nonlinear susceptibility of a metal surface is thought to arise from the free electrons in the surface region. Since free electrons are localised on formation of an adsorbate-substrate bond, adsorption leads to a quenching of the substrate nonlinearity. In this case A > B >> C and, from (16),

$$I(2\omega) \propto |A|^2 + 2|A||\Delta A|\Theta\cos\phi + |\Delta A\Theta|^2 \;, \tag{17}$$

where $\Delta A = B - A$, the difference between the substrate nonlinearity with and without adsorbate, and ϕ is the phase difference between the two signals. The generalisation of the result (17) to multiple sites has been discussed by Robinson and Richmond (1990). Equation (17) has been widely applied in the measurement of adsorption isotherms in UHV. The same model has been applied to the study of electrochemical interfaces. A mechanism for adsorption can be assumed and fitted to the data to obtain rate parameters and sticking coefficients. As in all isotherm studies, high quality data are required to distinguish between different mechanisms.

Adsorbate Spectroscopy. It is well known that the molecular polarisablility exhibits resonances when the applied electric field is of the same frequency as a molecular mode. The same is true for the hyperpolarisability. The most detailed description of the relationship between β and electronic structure is given by Ward (1965); in a simplified form for resonant SHG with a single electronic transition, we have

$$\beta_{ijk} = \frac{-e^3}{2\hbar^2}\left[\frac{\Delta r_n^i r_{ng}^j r_{ng}^k}{\omega_{ng}^2 - \omega^2} + r_{ng}^i\left(r_{ng}^j\Delta r_n^k + \Delta r_n^j r_{ng}^k\right)\frac{\omega_{ng}^2 + 2\omega^2}{\left(\omega_{ng}^2 - 4\omega^2\right)\left(\omega_{ng}^2 - \omega^2\right)}\right], \tag{18}$$

in which Δr_n is the dipole moment of the resonant excited state, r_{ng} is the transition dipole of the resonant transition and ω_{ng} the transition frequency. It is easy to see from (18) how resonance enhancement can make 1 or 2 elements of β dominant; for example if the dipole moment and transition dipole lie along the molecular z' axis, then $\beta_{z'z'z'}$ will be dominant. It is also noticeable from (18) that a molecule with a large excited state dipole moment and an intense transition will lead to a large nonlinear signal. These facts have been exploited in efforts to develop organic nonlinear optical materials.

An important application of surface nonlinear optical measurements is the determination of adsorbate vibrational spectra through the resonant enhancement of β. The emphasis has been on the application of infra-red (IR) + visible SFG to obtain adsorbate vibrational spectra (Hunt et al., 1987; Harris et al., 1989). A notable recent example was the first ever measurement of the vibrational spectrum of the liquid vapour interface (Superfine et al., 1991). In time domain measurements the resonance enhancement is a valuable means of studying selectively the vibrational relaxation kinetics of a resonant adsorbate.

The resonant part of the molecular hyperpolarisability (18) is written in condensed form

$$\beta_{i'j'k'} = \sum_q \frac{1}{\hbar}\frac{A_{k'}M_{i'j'}\Delta\rho}{\left(\omega_q - \omega_1 - i\Gamma_q\right)} \;, \tag{19}$$

where ω_q are the adsorbate vibrational frequencies (electronic resonances may be neglected here), ω_1 the frequency of the infra-red beam and $\Delta\rho$ the difference between ground (g) and the qth excited state population. The $A_{k'}$ and $M_{i'j'}$ are the infra-red transition dipole and Raman transition probability respectively. The condition for full resonance is $\omega_q = \omega_1$. The linewidth parameter Γ_q is introduced to prevent divergence. Clearly the selection rule $A_k \neq$

0, $M_{ij} \neq 0$ operates and the transition must be both IR and Raman allowed. The sum frequency spectrum is then recorded by scanning ω_1 through the vibrational frequencies of the adsorbate, and measuring the sum frequency intensity. A nice feature of this experiment is that not only are surface-specific spectra obtained, but vibrational signals are detected in the visible, where measurements with sensitive photomultiplier tubes are possible.

APPLICATIONS

The range of applications of SHG and SFG to the study of interfaces is ever increasing, as has been discussed in several reviews (Corn 1991, Corn and Higgins, 1994; Eisenthal 1992, 1993; Shen 1986, 1989; Meech, 1993). These applications include: the measurement of adsorbate orientation as a function of surface coverage or pressure at liquid/vapour, liquid/liquid, solid/liquid and electrochemical interfaces; the measurement of adsorbate vibrational spectra at various interfaces; measurement of the kinetics of interfacial chemical reactions. All of the above mentioned experiments can be tackled with the simple geometry of figure 2. With slightly more complex systems a variety of other experiments is possible: the adsorbate grating scattering experiment (Zhu, 1992) allows determination of the 2D diffusion coefficient; picosecond excited state reactions and ultrafast desorption kinetics can be studied with the picosecond pump and SHG probe methods (Sitzmann and Eisenthal, 1988; Meech and Yoshihara, 1989); adsorbate vibrational dynamics may be studied with picosecond time-resolved SFG methods (Harris et al.,1991); the homogeneous linewidth of adsorbates may be measured by IR photon echoes with an SFG 'reading' step (Guyot-Sionnest, 1991). The main point of this article has been to outline the principles underlying surface nonlinear optical experiments, but it seems appropriate to close with some illustrations of a few of the above mentioned applications.

Electrochemistry

In figure 5 the SH intensity generated at a silver electrode is measured as a function of time after a voltage step. The decrease in the SHG signal reveals, when the data are fitted to an equation of the form of (16), the kinetics of the electrode process. These data may be contrasted with standard electrochemical measurements: for more information see the reviews by Richmond et al. (1988) and Corn and Higgins (1994). Note that even with the simple set up of figure 2, millisecond kinetics can be recorded when the source of fundamental radiation is a quasi cw mode-locked laser.

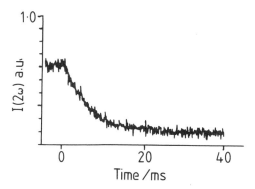

Figure 5. The kinetics of the double layer discharge and desorption of the perchlorate anion, observed in the time dependence of the SHG signal generated by a silver electrode following a potential step at time zero (adapted from Robinson and Richmond, 1990).

Adsorption Isotherms

In figure 6 the SH intensity generated at a copper (100) surface at 140 K is measured for different exposures of CO. The SHG signal arising from the bare metal is quenched by chemisorption, as described above. The data are fitted to equation (17) to obtain a coverage, which may be described by a Langmuir model to yield a sticking coefficient of 0.7.

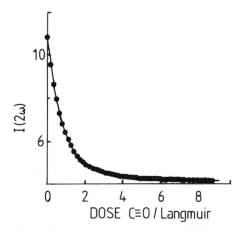

Figure 6. SHG from Cu (100) at 140 K, measured as a function of CO exposure. The data are fitted to a Langmuir model (adapted from Shen, 1986).

Chemical Reactions

An example of the use of the SHG signal to measure reaction kinetics is shown in figure 7. First the SHG signal from Si(111) for different coverages of H was measured to provide a calibration curve (the dependence on coverage for this system is not a simple one). Next the SH intensity of a Si(111)H surface was monitored as a function of time after a temperature jump. Conversion of the measured intensity to a coverage using the calibration data then yields the kinetics of the recombinative desorption reaction. Both mechanism and energetic information are thereby obtained.

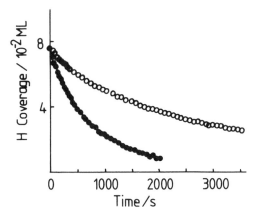

Figure 7. Kinetics of recombinative desorption of H_2 from a silicon-H surface. The SHG signal, measured as a function of time following a temperature jump, was converted to coverage using an earlier calibration. (Adapted from Reider et al., 1991).

Ultrafast Kinetics

Using a slightly more complex optical layout it is possible to record the picosecond kinetics of adsorbates. The layout for the ultrafast pump-SHG probe experiment is shown in figure 8. Essentially the pump induces some perturbation to the surface, and the probe generated SHG, measured as a function of the delay time Δt, monitors the surface relaxation. In figure 9 the picosecond kinetics of the dye malachite green on silica are shown. Malachite green is

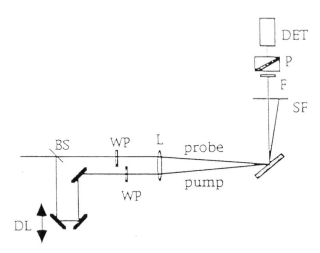

Figure 8. Experimental arrangement for a picosecond time resolved SHG experiment. The probe-generated SHG signal is recorded for different pump-probe delays, selected with the delay line, DL.

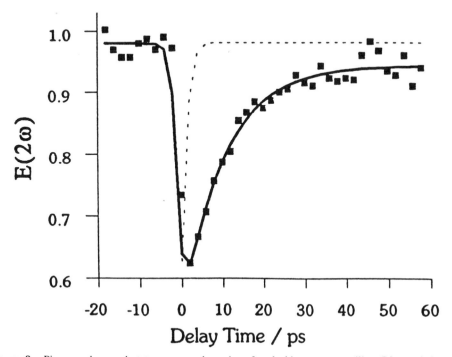

Figure 9. Picosecond ground state recovery dynamics of malachite green on silica (Morgenthaler and Meech, 1992)

known to undergo a fast conformational change in the excited state leading to rapid repopulation of the ground state. Evidently the reaction also occurs in the adsorbate. In figure 9 there is an instantaneous decrease in the SHG signal at t = 0. This arises from the pump pulse induced depletion of the resonant ground state. As the excited state reaction proceeds the ground state is repopulated leading to a recovery in the SHG signal. The rate of recovery may be related to the rate of reaction (Morgenthaler and Meech, 1992). Ultrafast vibrational relaxation in adsorbates can be measured in a similar way using an IR pump - SFG probe method (Harris et al., 1991).

SUMMARY

Second order nonlinear optical signals provide unique information about surface phenomena. The technique is applicable to a range of interfaces that cannot be studied by conventional surface spectroscopic methods. To date, however, relatively few interfacial systems have been studied in detail. It is hoped that the preceding description of the fundamental and applied aspects of surface nonlinear optical signals will interest others in applying these techniques to interfacial systems.

ACKNOWLEDGEMENTS

MJEM is grateful to the Carnegie Trust for the Universities of Scotland for a scholarship. SRM acknowledges the financial support of the SERC.

REFERENCES

Andrews, D.L., 1993, *J. Mod. Optics* 40:939.
Bloembergen, N., and Pershan, P.S., 1962, *Phys. Rev.* 128:606.
Corn, R.M., 1991, *Anal. Chem.* 63:4285.
Corn, R.M., and Higgins, D.A., 1994, *Chem. Rev.* 94:107.
Eisenthal, K.B., 1993, *Acc. Chem. Res.* 26:636.
Eisenthal, K.B., 1992, *Ann. Rev. Phys. Chem.* 43:627.
Goldstein, H., 1969, "Classical Mechanics," sixth edition, Addison-Wesley, New York.
Grudzkov, Y.A., and Parmon, V.N., 1993, *J. Chem. Soc. Faraday Trans.* 89:4017.
Guyot-Sionnest, P., 1991, *Phys. Rev. Lett.* 66:1489.
Harris, A.L., Chidsey, C.E.D., Levinos, N.J., and Loicano, D.N., 1989, *Chem. Phys. Lett.* 141:350.
Harris, A.L., Rothberg, L., Dhar, L., Levinos, N.J., and Dubois, L.H., 1991, *J. Chem. Phys.* 94:2438.
Heinz, T.F., Loy, M.M.T., and Thomson, W.A., 1985, *Phys. Rev. Lett.* 54:63.
Heinz, T.F., 1991, Second order nonlinear optical effects at surfaces and interfaces, *in*: "Nonlinear Electromagnetic Phenomena," H.E.Ponath and G.E. Stegeman, eds., Elsevier, Amsterdam.
Hunt, J.H., Guyot-Sionnest, P., and Shen, Y.R., 1987, *Chem. Phys. Lett.* 133:189.
Jenkins, F.A., and White, H.F., 1981, "Fundamentals of Optics," fourth edition, McGraw-Hill, New York.
Meech, S.R., and Yoshohara, K., 1989, *Chem. Phys. Lett.* 154:20.
Meech, S.R., 1993, Kinetic applications of surface nonlinear optical signals, *in*: "Advances in Multiphoton Processes and Spectroscopy," S.H. Lin, A. Villaey and Y. Fujimura, eds., World Scientific, Singapore.
Mizrahi, V., and Sipe, J.E., 1988, *J. Opt. Soc. Am.* B5:660.
Morgenthaler, M.J.E., and Meech, S.R., 1992, Ultrafast torsional dynamics in adsorbates, *in*: "Ultrafast Phenomena VIII," J-L. Martin et al., eds., Springer Series in Chemical Physics 55:606.
Richmond, G.L., Robinson, J.M., and Shannon, V.L., 1988, *Progress in Surface Science* 29:1.
Robinson, J.M., and Richmond, G.L., 1990, *Chem. Phys.* 141:175.
Sitzmann, E.V., and Eisenthal, K.B., 1988, *J. Phys. Chem.* 92:4579.
Shen, Y.R., 1986, *Ann. Rev. Mater. Sci.* 16:69.
Shen, Y.R., 1989, *Ann. Rev. Phys. Chem.* 40:327.

Superfine, R., Huang, J.Y., and Shen, Y.R., 1991, *Phys. Rev. Lett.* 66:1066.

Ward, J.F., 1965, *Rev. Mod. Phys.* 37:1.

Zhang, T.G., Zhang, C.H., and Wong, G.K., 1990, *J. Opt. Soc. Am.* B7:902.

Zhu, X.H., 1992, *Mod. Phys. Lett.* B6:1217.

LASER DESORPTION

Anita C. Jones

Department of Chemistry
University of Edinburgh
West Mains Road
Edinburgh, EH9 3JJ

INTRODUCTION

Strictly, desorption is the term used to describe the transfer of atoms or molecules from a surface-adsorbed state into the gas phase. The study of pure desorption processes lies in the realm of surface science and is carried out using a well-defined sample, consisting of a submonolayer of atoms or small molecules, adsorbed to a surface of known structure. Mass spectrometrists use the term 'desorption' in a less rigorous sense to describe a range of techniques which are used to transfer atoms or molecules from the solid state into the gas phase. These methods include field desorption, plasma desorption, secondary ion mass spectrometry, fast atom bombardment and laser desorption. The sample is typically many monolayers thick, deposited on a crude substrate, and may consist of a mixture of complex organic molecules, one component of which acts as a matrix. In these methods, the desorption event generally results in the removal of several monolayers of material.

Laser desorption may be defined as the use of short, intense pulses of laser light to induce the formation of intact gaseous molecular ions or neutral molecules. One of the most important features of this desorption technique is that it enables the vaporisation of thermally labile species, such as biomolecules, without decomposition. The use of laser desorption for the vaporisation of involatile, thermally labile organic molecules was first demonstrated by Kistemaker's group (Posthumus et al., 1978). They showed that pulsed CO_2 laser radiation could be used to generate gaseous cationised species of intact biomolecules, such as oligosaccharides and oligopeptides. Since then, laser desorption has developed in two distinct directions: (i) matrix-assisted laser desorption/ionisation (MALDI), in which pulsed ultraviolet radiation is used to produce intact gaseous *molecular ions*, and (ii) laser desorption of intact *neutral molecules*, which is generally achieved using pulsed infrared radiation from a CO_2 laser. The practical methodology, mechanism and applications of each of these techniques will be dealt with in this chapter. At present, MALDI is by far the more widely used technique because of its extremely successful application to the mass spectrometry of high molecular weight biomolecules. The considerable potential of infrared laser desorption has yet to be fully exploited.

An Introduction to Laser Spectroscopy
Edited by David L. Andrews and Andrey A. Demidov, Plenum Press, New York, 1995

MATRIX-ASSISTED LASER DESORPTION/IONISATION

Matrix-assisted laser desorption/ionisation (MALDI) is a combined desorption and ionisation technique. Pulsed ultraviolet laser radiation is used to produce intact gaseous molecular or quasi-molecular ions from sample molecules finely dispersed in a solid matrix of small organic molecules that absorb strongly at the desorption laser wavelength. This method of laser desorption forms the basis of MALDI mass spectrometry, which is now widely used for the mass analysis of proteins with molecular masses in the range of one thousand to several hundred thousand Daltons.

The principle of matrix-assisted laser desorption for large organic molecules was first demonstrated by Tanaka et al. (1988). They used a liquid matrix, in which was suspended a fine metal powder, and produced mass spectra of proteins of molecular masses up to 35 kDa. Karas and Hillenkamp (1988) were first to demonstrate the use of a solid matrix, and went on to carry out the pioneering work which paved the way for the development of MALDI to its current position as a standard technique for the mass analysis of high molecular weight species. Beavis and Chait (1989) also made important contributions to the early development of the technique, particularly in the development of new matrices. These four workers and their groups have continued to dominate the advancement of MALDI mass spectrometry; they have published a number of excellent reviews on the subject, for example Beavis (1992), Hillenkamp et al. (1991) and Bahr et al. (1994).

PRACTICAL ASPECTS OF THE MALDI TECHNIQUE

In practical terms, there are two key aspects of the MALDI technique to be considered: the laser source, which provides the energy for the desorption/ionisation process, and the method of sample preparation.

Laser Properties

The type of laser most commonly used for matrix-assisted laser desorption is the nitrogen (N_2) laser, which has an emission wavelength of 337 nm. The 355 nm (frequency-tripled) and 266 nm (frequency-quadrupled) outputs of the Nd:YAG laser are also used. Typically, laser pulse durations in the range 1 - 200 ns are used, with irradiances of 10^6 to 10^7 Wcm^{-2}. These power densities are achieved by focusing the laser beam to a spot of 30 - 500 µm diameter on the sample surface. The irradiance at the sample surface is a critical parameter for achieving successful desorption/ionisation. There is a well-defined threshold (minimum) irradiance for the production of ions, and the best results appear to be obtained at irradiances not more than about 20% above the threshold value

Sample Preparation

The analyte molecules, which are most often some type of biopolymer, are dispersed in a matrix of small organic molecules which have a strong optical absorption at the desorption laser wavelength. The use of a strongly uv-absorbing matrix means that desorption and ionisation of the analyte molecules can be achieved regardless of the latters' own absorption properties. The sample is prepared by mixing a small volume, typically 0.5 - 10 µl, of concentrated matrix solution (5 - 10 g l^{-1} in H_2O or a mixed aqueous/organic solvent) with a similar volume of dilute analyte solution (typically, 10^{-5}-10^{-7}M) on a stainless steel sample probe. The solvent is then evaporated and the sample transferred to the vacuum chamber of the mass spectrometer. The incorporation of the analyte molecules into the crystalline matrix is essential for the production of intense ion signals in the MALDI process.

Selection of a suitable matrix and careful sample preparation are crucial for successful laser desorption/ionisation. The choice of matrix compound depends, of course, on the laser wavelength; it also depends on the solubility of the analyte, since the matrix solution and analyte solution must be miscible, and on the class of compound to which the analyte belongs. For example, 2,5,-dihydroxybenzoic acid and synapinic acid are matrices commonly used with 337 nm or 355 nm laser wavelengths for MALDI of proteins, while 3-hydroxy picolinic acid has been found to be particularly suitable for oligonucleotides (Wu et al.,1993). An extensive study of potential matrices has been reported by Beavis (1992).

THE MECHANISM OF MALDI

The mechanisms involved in MALDI remain poorly understood and the technique has been developed empirically. A detailed discussion of the possible MALDI mechanisms may be found in Vertes and Gijbels (1993). A brief, qualitative description of the mechanistic role of the matrix in the desorption/ionisation process is given here. The matrix can be considered to perform three major functions (Bahr et al., 1994):

(i) The matrix absorbs the laser radiation. The absorption of the ultraviolet laser radiation by the matrix molecules results in the rapid deposition of a large amount of energy into the solid matrix-analyte solution. This leads to an explosive disintegration of a small matrix-analyte volume and the release of analyte (and matrix) molecules into the vapour phase, with little internal excitation. A model for desorption of molecules without thermal decomposition will be presented below, when we consider infrared laser desorption.

(ii) The matrix isolates the analyte molecules. The large molar and volume excess of matrix molecules separates the analyte molecules from each other, thus reducing the strong intermolecular interactions between them, facilitating the desorption of individual molecules rather than aggregates.

(iii) The matrix is involved in the ionisation of the analyte molecules. The ions produced in the MALDI process are generally quasi-molecular ions, rather than radical cations. These quasi-molecular ions are cationised species, such as protonated ($M-H^+$) or alkalinated (e.g. $M-Na^+$) molecular ions. Little is known about the formation of these ions. The matrix is thought to perform an active role in protonation, by photoexcitation or photoionisation of the matrix molecules followed by proton transfer to the analyte molecules. Alkalinated ions are probably the products of gas phase ion-molecule reactions.

MALDI MASS SPECTROMETRY

MALDI mass spectrometry is achieved by combining MALDI with time-of-flight mass spectrometry. The short pulse of ions produced by laser desorption is ideally suited to mass analysis by the time-of-flight technique which makes it possible to record a complete mass spectrum for each laser shot. The use of MALDI mass spectrometry has spread rapidly and the availability of a number of commercial instruments means that it is now a standard technique for the mass analysis of biopolymers, particularly proteins. A number of comprehensive, recent reviews of MALDI mass spectrometry can be found in the literature, for example: Bahr et al. (1994), Burlinghame et al. (1994), Vertes and Gijbels (1993), Sundqvist (1992). A detailed account of MALDI mass spectrometry will not be given here, therefore, but the main applications of the technique are summarised below (see also the chapter by Ledingham).

Currently, the primary application of MALDI mass spectrometry is measurement of the molecular masses of proteins. The mass spectra of hundreds of different proteins, with molecular masses up to 300,000 Da, have been determined. The applicability of MALDI to proteins appears to be independent of the primary, secondary or tertiary structure of the molecules, and glycoproteins containing large proportions of carbohydrate can also be analysed. A great strength of the MALDI technique is its sensitivity. Typically 1 picomole of analyte is loaded on to the sample probe; most of this material is not consumed and can be recovered after the analysis. In some cases, as little as 1-10 femtomoles of protein may be sufficient for analysis (e.g. Karas et al., 1989; Strupat et al., 1991). Another important feature of MALDI is its high tolerance to inorganic and organic contaminants. This means that buffers and salts normally present in biochemical samples do not have to be removed before mass analysis.

In MALDI mass spectra of proteins, the most intense signal is, generally, the singly charged quasi-molecular ion which is formed by protonation of the molecule. Doubly and triply charged quasi-molecular ions as well as singly and multiply charged cluster ions may also be observed. Fragmentation due to cleavage of the bonds of the protein backbone is not generally observed under normal conditions and MALDI does not, therefore, provide information on protein structure, only molecular mass. The MALDI mass analysis of peptide mixtures produced by enzymatic or chemical digestion of proteins is now being developed as a new strategy for the elucidation of protein structure (see for example, Billeci and Stults, 1993). Other recent developments in the application of MALDI to protein analysis include the combination of MALDI with separation techniques such as gel electrophoresis (Eckerskorn et al., 1992) and capillary zone electrophoresis (Keough et al., 1992; van Veelen et al., 1993).

MALDI mass spectrometry can also be applied to other biomolecules, such as oligonucleotides and oligosaccharides. Oligo(deoxy)ribonucleotides (e.g. Wu et al., 1993) and small RNA molecules (e.g. Nordhoff et al., 1992) have been analysed. So far, the highest mass detected for a laser-desorbed nucleic acid ion is about 40,000 Da. A variety of oligosaccharide species have been successfully analysed; see for example Stahl et al. (1991), Juhasz and Costello (1992) and Harvey (1993). Large oligosaccharides appear to be difficult to ionise and the highest mass detected to date is only about 15,000 Da.

Recently, the use of MALDI mass spectrometry has been extended to synthetic polymers. Polar polymers, which are miscible with the well known matrices, are amenable to the MALDI technique. For example, poly(ethylene glycols) up to 40 kDa (Bahr et al., 1992) and poly(styrene sulfonic acid) of average molecular mass 200 kDa (Danis et al., 1992) have been analysed successfully. In order to apply MALDI to the larger and more important class of non-polar synthetic polymers, new matrices must be devised. For example, Bahr et al. (1992) have recorded mass spectra of polystyrenes up to 70 kDa using nitrophenyl octyl ether, a highly viscous liquid, as a matrix. An alternative approach, which has been applied recently to polystyrene and polybutadiene, is to use no matrix but to dope the sample with metal salts to induce cationisation (Mowat and Donovan, 1995).

LASER DESORPTION OF NEUTRAL MOLECULES

Early experiments on pulsed CO_2 laser desorption were concerned with the direct generation of ions, rather than neutral molecules. However, Cotter and co-workers, (Cotter, 1980; van Bremen et al., 1983) showed that neutrals are produced in addition to ions and, furthermore, that the yield of neutrals is much greater than that of ions. Typically the ratio of neutrals to ions is $\geq 10^4:1$, for power densities $< 10^8$ W cm^{-2}. The formation of ions and neutrals occurs in two separate processes that are temporally distinct. The production of ions occurs promptly and persists for about 1 μs after the laser pulse, whereas the production of neutral molecules persists for about 100 μs.

The first use of laser desorption for the deliberate and specific purpose of generating gas phase neutral molecules, rather than ionic species, may be attributed to Schlag and co-workers (Henke et al., 1983). By subjecting the desorbed neutrals to laser multiphoton post-ionisation, followed by time-of-flight mass analysis, they showed that pulsed CO_2 laser desorption could achieve the vaporisation of thermally labile organic molecules as intact neutrals species, without thermal decomposition. Following these initial experiments, laser desorption of neutral molecules has been developed primarily in combination with laser photoionisation as a mass spectrometric technique. The applications of laser desorption are not, however, limited to mass spectrometry. Laser desorption can be used as a source of gas phase molecules for supersonic jet spectroscopy, thus enabling the high resolution vibronic spectroscopy of thermally labile biomolecules. For example, Levy's group (Cable et al., 1988, 1989) have used this technique to study small, tryptophan-containing peptides. Another application of laser desorption is as a sampling or separation technique to transfer an analyte from its native substrate to another substrate for analysis. For example, de Vries et al. (1990) have demonstrated the 'laser transfer' of organic surface films for infrared analysis.

Laser desorption of neutral molecules can, in fact, be achieved with pulsed infrared, ultraviolet or visible laser radiation. We shall concentrate here on pulsed infrared desorption by CO_2 laser radiation of wavelength 10.6 μm, which is the most commonly used technique.

PRACTICAL ASPECTS OF INFRARED LASER DESORPTION

As in the case of MALDI, we need to consider the properties of the laser used to achieve desorption and the manner in which the sample is prepared and presented to the desorption laser beam.

Laser Properties

A pulsed TEA CO_2 laser of wavelength 10.6 μm is used, with a pulse duration of ca 100 ns. The infrared laser radiation is focused on the sample to a spot typically of diameter 100 μm - 1 mm, although for spatially resolved studies (see below) spot sizes of 20 - 50 μm have been used. The irradiances used to achieve desorption of neutral molecules are generally in the range of 10^6-10^8 W cm^{-2}. However, unlike MALDI, the irradiance is not a critical parameter. There is a well-defined threshold, above which successful desorption of neutrals can be achieved over a wide irradiance range (often several orders of magnitude). At high power densities ($>10^8$ W cm^{-2}) a significant yield of ions is produced in addition to neutrals.

Sample Presentation

It is in the manner of sample preparation that pulsed infrared desorption differs radically from MALDI. No matrix is required and, furthermore, no special sample preparation of any kind is necessary. The sample can be presented to the desorption laser in variety of ways. Most commonly, the sample is presented as a solid layer, deposited from solution, on a substrate which absorbs at the 10.6 μm laser wavelength. Typical substrates are stainless steel, glass and macor. Alternatively the sample may be presented as a suspension in a viscous fluid such as glycerol, deposited on an infrared-absorbing substrate, or as a solid pellet. The versatility of infrared laser desorption with respect to sample presentation is one of its great strengths. As discussed below, this technique can be used to desorb analyte molecules directly from 'real' complex samples such as contaminated soil.

THE MECHANISM OF INFRARED LASER DESORPTION

Infrared laser desorption is essentially a laser-induced thermal desorption process. At present the mechanism of this process, whereby thermally labile molecules can be vaporised without decomposition, is not fully understood. In the case of a thin (essentially infrared-transparent) sample layer deposited on an infrared-absorbing substrate, it is generally accepted that the infrared laser pulse is absorbed by the substrate, resulting in rapid local heating ($>10^8$ K s^{-1}) of the substrate surface and subsequent desorption of the sample molecules. Zare and Levine (1987) have proposed a model to account for the desorption of internally cool molecules in this type of rapid thermal desorption process. In the case of thick samples, the generation of shock waves and mechanical stress in the sample may be involved in the desorption mechanism.

Rapid Thermal Desorption

The energy 'bottleneck' mechanism proposed by Zare and Levine (1987) to account for the laser-induced thermal desorption of internally cool molecules can be summarised qualitatively as follows. A molecule physisorbed to the substrate surface is bound by a weak, van der Waals-type bond. This bond will have a low vibrational frequency, similar in magnitude to those of the surface phonons of the substrate. However, the chemical bonds of the physisorbed molecule will have much higher vibrational frequencies. When the substrate is heated as a result of absorption of the laser radiation, energy will rapidly be transferred from the surface phonons to the surface-adsorbate physisorption bond. But the energy flow from the surface-adsorbate bond into the chemical bonds of the adsorbed molecule will be slow, because of the mismatch in vibrational frequencies. The physisorption bond thus acts as a bottleneck for energy transfer from the rapidly heated surface to the internal degrees of freedom of the adsorbed molecule, so that desorption occurs without the molecule becoming internally hot. Provided the rate of heating of the substrate, and hence the rate of energy flow into the physisorption bond, is sufficiently high, the adsorbed molecule will desorb without thermal decomposition.

Shock-Wave and Mechanical Desorption

For thick samples, a shock-wave driven desorption mechanism has been suggested (Lindner and Seydel, 1985). It is proposed that thermal desorption at the substrate-sample interface results in a shock wave which traverses the sample layer leading to desorption of intact molecules from the sample surface. A similar phenomenon which may be involved in the desorption of thick samples is that of disintegration via mechanical stress (Beavis et al., 1988). Here, laser-induced inhomogeneous heating of the sample produces thermal expansion, resulting in mechanical stress and disintegration of the sample with the ejection of molecules into the vapour phase.

LASER-DESORPTION LASER-PHOTOIONISATION TIME-OF-FLIGHT MASS SPECTROMETRY (L²TOFMS)

The primary application of infrared laser desorption is, in combination with laser photoionisation, as a mass spectrometric technique. The laser-desorbed neutral molecules are multiphoton-ionised by ultraviolet laser radiation and the resulting ions are mass-analysed by time-of-flight mass spectrometry. This technique is known as laser-desorption laser-photoionisation time-of-flight mass spectrometry (L²TOFMS) or, sometimes, simply two-step laser mass spectrometry. Early advances in this area were dominated by three groups: Schlag, Grotemeyer et al. in Munich (Weyssenhof et al., 1985; Grotemeyer et al., 1986);

Lubman et al. in Michigan (Tembreull and Lubman, 1986; Lubman, 1987) and Zare et al. in Stanford (Engelke et al., 1987; Hahn et al., 1987). A number of other groups (including de Vries et al. at IBM, San Jose; Langridge-Smith et al. in Edinburgh and Zenobi et al. in Zürich) has subsequently become involved in the development of L^{2}TOFMS and its applications.

L^{2}TOFMS has been shown to be applicable to a wide range of involatile and/or thermally unstable organic and bio-organic molecules, including polycyclic aromatic compounds, dyestuffs, porphyrins, amino acids, peptides, vitamins, nucleic bases and nucleotides. Recently, there has been increasing interest in the use of L^{2}TOFMS as an analytical technique. A number of reviews of L^{2}TOFMS and its applications may be found in the literature, e.g. Grotemeyer and Schlag (1989); Lubman (1993), Kovalenko et al. (1991) and Zenobi (1994). The account given here will be limited to a brief description of the methodology of the technique and a consideration of its analytical potential.

Instrumentation and Experimental Methodology

The Edinburgh L^{2}TOFMS instrument is shown schematically in figure 1. The mass spectrometer essentially consists of two differentially pumped high vacuum chambers: the ionisation chamber and the chamber housing the reflectron time-of-flight mass analyser.

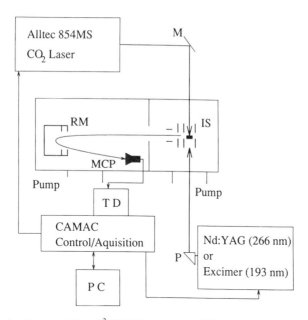

Figure 1. Schematic diagram of the L^{2}TOFMS instrument: IS, ion source region; M, mirror, MCP, microchannel plate detector; P, right angle prism; PC, personal computer; RM reflectron mirror; TD, transient digitiser.

The sample is placed in the ionisation chamber of the mass spectrometer, as shown in figure 2. The sample is mounted on an xyz manipulator so that it can be moved relative to the CO_2 laser beam, allowing the full area of the sample to be explored by the desorption laser spot. The desorbed pulse of neutral molecules is intersected approximately 4 mm above the sample surface by a pulse of ultraviolet radiation, from a Nd:YAG laser (266 nm) or an excimer laser (193 nm), causing two-photon ionisation. The time delay between the firing of the desorption laser and the photoionisation laser is adjusted to optimise the molecular ion signal. The photoions are mass separated in the reflectron time-of-flight mass analyser and detected by a microchannel plate detector. Mass spectra are collected using a CAMAC-based data acquisition system, employing a 200 MHz transient digitiser, interfaced to a personal computer. Data from a number of laser shots, typically 10 to 200 depending on signal intensity and stability, are accumulated for each mass spectrum. The complete experimental procedure, from sample presentation to data accumulation and storage, can be performed within about 10 minutes.

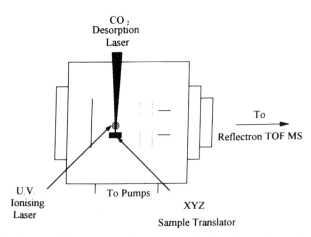

Figure 2. Schematic diagram of the experimental arrangement for laser desorption and laser photoionisation.

L²TOFMS as an Analytical Technique

L²TOFMS has a number of advantageous features which make it a very effective technique for the analysis of complex 'real-world' samples. Many of these advantages arise from the spatial and temporal separation of the desorption and ionisation steps which enables them to be optimised independently, with respect both to laser wavelength and power density. The application of infrared laser desorption to multicomponent mixtures of compounds allows simultaneous vaporisation of the various components as intact neutral molecules. The photoionisation laser wavelength and power density can then be selected to give so-called 'soft' ionisation which occurs with minimal fragmentation. This results in simple, readily interpretable mass spectra consisting principally (often exclusively) of molecular ion peaks. By increasing the photoionisation power density, 'hard' ionisation conditions can be produced in order to induce fragmentation and provide structural information. For example, Grotemeyer and Schlag (1988) have demonstrated the use of L²TOFMS to determine the amino acid sequence of small peptides.

The separation of the desorption and ionisation steps also allows us to exploit the ability of resonance-enhanced two-photon ionisation (R2PI) to ionise selectively target molecules that have appreciable absorption cross-sections at a particular laser wavelength. This wavelength-selectivity of R2PI is particularly valuable in minimising matrix interference in the mass spectra of complex samples. Furthermore, the use of R2PI in combination with pulsed laser desorption can provide very high detection sensitivities: detection limits as low as zeptomoles (10^{-21} moles) have been reported for some species (Maechling et al., 1993). Another important advantage of L²TOFMS is that infrared laser desorption is not prone to matrix effects; it can be applied to a wide variety of sample types and precludes the need for time-consuming extraction and pre-separation procedures.

194

The L²TOFMS technique does, of course, have limitations and disadvantages. The analyte must contain an ultraviolet chromophore in order to achieve efficient resonance-enhanced two-photon ionisation. This limitation can be overcome by the use of vacuum ultraviolet laser radiation to achieve single-photon ionisation (see, for example, Koster and Grotemeyer, 1992). However, this adds considerably to the complexity of the technique. A major limitation of L²TOFMS is the difficulty in performing quantitative measurements. A fundamental problem in attempting to carry out absolute quantitation is that reliable calibration experiments are difficult to perform. However, it is possible to determine a lower limit of detection for a particular species. It is important to be aware that absolute detection limits are both molecule- and wavelength-specific. This means that an L²TOF mass spectrum does not give direct information on the relative concentrations of components in a mixture, unless they all have very similar photoionisation cross-sections at the wavelength used. A further disavantage arises from the decrease in photoionisation efficiency with increasing molecular size (Schlag et al., 1992). This means that, unlike MALDI which can access very high molecular masses, L²TOFMS is limited to relatively low molecular masses, less than 10,000 Da. It must also be acknowlegded that, whatever the merits of L²TOFMS, the experimental complexity and instrumentation costs of a two-laser system are bound to be a barrier to its widespread adoption as an analytical technique.

The use of L²TOFMS for the direct analysis of target species in complex matrices is illustrated well by some work which we have recently carried out on the determination of polycyclic aromatic hydrocarbons (PAH's) in contaminated soils (Dale et al., 1993, 1994). The soil samples were presented directly to the mass spectrometer, as received, without pre-separation or chemical treatment. The soil was simply ground to a fine powder, mixed to a paste with a drop of glycerol and applied to the sample probe. Typical mass spectra obtained

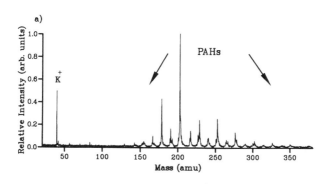

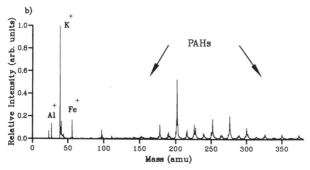

Figure 3. L²TOF mass spectrum of ca 1mg of contaminated soil obtained using (a) 266 nm and (b) 193 nm laser photoionisation.

for one of these samples, using 266 nm and 193 nm photoionisation, are shown in figure 3 (a) and (b), respectively. Each mass spectrum was recorded as the sum of 200 laser shots and represents the total ion signal obtained on interrogating approximately 1 mg of the contaminated soil. A striking feature of these spectra is the absence of peaks due to the soil matrix itself or the glycerol binder. Signals due to the PAH contaminants can clearly be seen in the region between 100 and 400 Da. Signals at masses higher than 300 Da indicate the presence of PAH's with more than seven fused rings. In the low mass region, below 100 Da, the dominant signals are due to elemental cations, e.g K^+, Al^+, and Fe^+. These mass spectra were recorded under soft photoionisation conditions. Therefore, they are free from fragment ions, allowing the molecular ions of the PAH's present in the sample easily to be identified.

The soil samples had been investigated previously using gas chromatography-mass spectrometry (GC-MS), following soxhlet extraction with dichloromethane, and a number of specific PAH's had been identified. It was possible to assign these previously identified species to a number of the intense peaks in the L^2TOF mass spectra, although isomers cannot be differentiated. For example, peaks at 178, 202, 28, 252, and 276 Da correspond to phenanthrene/anthracene, fluoranthene/pyrene, chrysene/benz[a]anthracene, benz[b or k]-fluoranthene/benzo[a]pyrene and indeno-[123-cd]pyrene/benzo[ghi]perylene, respectively. The peaks corresponding to these species are numbered in figure 4, which shows an expansion of the PAH-containing region of the mass spectrum taken using 193 nm photoionisation. In the previous GC-MS work, the various PAH's had been quantifiably determined to be present in concentrations between about 0.1 ppm and several hundred ppm. To assess the feasibility of quantifying the concentration of PAH's using L^2TOFMS, the concentration of phenanthrene/anthracene (peak 1 in figure 4) was determined using standard additions of phenanthrene. The concentration of phenanthrene/anthracene was estimated to be 390 ppm, in reasonable agreement with the combined concentration of 310 ppm for anthracene and phenanthrene obtained by GC-MS.

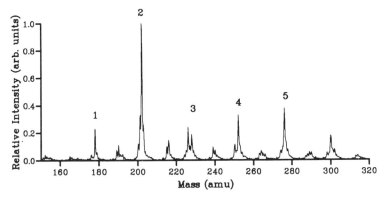

Figure 4. Expansion of L^2TOF mass spectrum of contaminated soil, between 170 and 320 Da, to show more clearly the resolved mass spectral peaks. Mass assignments are: 1, 178, phenanthrene/anthracene; 2, 202, pyrene/fluoranthene; 3, 228, chrysene/benz[a]anthracene; 4, 252, benzo[a]fluoranthene/ benzo[a]pyrene; 5, 276, indeno[123-cd]pyrene/benzo[ghi]perylene.

The work described above demonstrates the capability of L^2TOFMS to directly screen environmental samples for PAH's, in a rapid, selective and sensitive manner. L^2TOFMS has also been applied successfully to the direct detection of PAH's in engine oil, coal soot and coal-tar creosote (Dale et al., 1994); in atmospheric aerosol particulates (Zenobi, 1994; Dale et al., 1995) and in meteorites (Kovalenko et al.,1992). The range of applications of L^2TOFMS is steadily increasing, with an expanding range of molecular species being examined and technical advances in instrumentation being made. An important new development is the introduction of laser mass microscopy. This exploits the high sensitivity of laser photoionisation time-of-flight mass spectrometry and the high spatial resolution which can be achieved by tightly fosusing the desorption laser beam. Spatial resolutions of a few tens of micrometers have been achieved using infrared laser desorption (Kovalenko et al., 1992; Voumard et al., 1993), while a resolution of 1 µm has been achieved using an ultraviolet waveguide excimer laser for desorption (de Vries et al.,1992). This capacity for high spatial resolution gives L^2TOFMS the potential for new and important microanalytical tasks, such as the two-dimensional mapping of molecular adsorbates on a wide range of substrates, the examination of individual particles and, perhaps, even individual cells.

ACKNOWLEDGEMENTS

I gratefully acknowledge my co-workers in the Edinburgh L^2TOFMS group, particularly Pat Langridge-Smith and Mike Dale, whose work I have drawn upon in writing this article. I am also grateful to Ian Mowat for informative discussions on MALDI mass spectrometry.

REFERENCES

Bahr, U., Deppe, A., Karas, M., Hillenkamp, F., and Giessman, U., 1992, *Anal. Chem.* 64:2866.
Bahr, U., Karas, M., and Hillenkamp, F., 1994, *J. Anal.Chem.* 348:783.
Beavis, R.C., Lindner, J., Grotemeyer, J., and Schlag, E.W.,1988, *Z. Naturforsch* 43a:1083.
Beavis, R.C., and Chait, B.T., 1989, *Rapid Commun. Mass Spectrom.* 3:233.
Beavis, R.C., 1992, *Org. Mass. Spectrom.* 27:653.
Billeci, T.M., and Stults, J.T., 1993, *Anal.Chem.* 65:1709..
Burlinghame, A.L., Boyd, R.K., and Gaskell, S.J., 1994, *Anal.Chem.* 66:634R.
Cable, J.R., Tubergen, M.J., and Levy, D.H., 1988, *J. Am. Chem. Soc.* 110:7349.
Cable, J.R., Tubergen, M.J., and Levy, D.H., 1989, *J. Am. Chem. Soc.* 111:9032.
Cotter, R.J., 1980, *Anal. Chem.* 52:1767.
Dale, M.J., Jones, A.C., Pollard, S.J.T., Rowley, A.G., and Langridge-Smith, P.R.R., 1993, *Env. Sci. Technol.* 27:1693.
Dale, M.J., Jones, A.C., Pollard, S.J.T., and Langridge-Smith, P.R.R., 1994, *Analyst* 119:571.
Dale, M.J., Downs, O.H.J., Costello, K.F., Wright, S.J., Langridge-Smith, P.R.R., and Cape, N.J., 1995, *Environ. Pollut.* (in press).
Danis, P.O., Karr, D.E., Mayer, F., Holle, A., and Watson, C.H., 1992, *Org. Mass Spectrom.* 27:843.
De Vries, M.S., Hunziker, H.E., Wendt, H.R., and Saperstein, D.D.,1990, *Anal. Chem.* 62:2385.
Eckerskorn, C., Strupart, K., Karas, M., Hillenkamp, F., and Lottspeich, F., 1992, *Electrophoresis* 13:664.
Engelke, F., Hahn, J.H., Henke, W., and Zare, R.N., 1987, *Anal. Chem.* 59:909.
Grotemeyer, J., Boesl, U., Walter, K., and Schlag, E.W., 1986, *Org. Mass Spectrom.*, 21:645.
Grotemeyer, J., and Schlag, E.W., 1988, *Org. Mass Spectrom.* 23:388.
Grotemeyer, J., and Schlag, E.W., 1989, *Acc. Chem. Res.* 22:399.
Hahn, J.H., Zenobi, R., and Zare, R.N., 1987, *J. Am. Chem. Soc.* 109:2842.
Henke, W.E., Weyssenhoff, H.V., Selzle, H.L., and Schlag, E.W., 1983, *Verh. Dtsch. Phys. Ges.* 3:139.
Hillenkamp, F., Karas, M., Beavis, R.C., and Chait, B.T., 1991, *Anal. Chem.* 63:1193A.
Karas, M., and Hillenkamp, F., 1988, *Anal. Chem.* 60:2299.
Karas, M., Ingendoh, A., Bahr, U., and Hillenkamp, F., 1989, *Biomed. Environ. Mass Spectrom.* 18:841.
Keough, T., Takigiku, R. Lacey, M.P., and Purdon, M., 1992, *Anal. Chem.* 64:1594.
Koster, C., and Grotemeyer, J., 1992, *Org. Mass Spectrom.*27:463.
Kovalenko, L.J., Philippoz, J.M., Bucenell, J.R., Zenobi, R., and Zare, R.N., 1991, *Space Science Reviews* 56:191.
Lindner, B., and Seydel, U., 1985, *Anal. Chem.* 57:895.
Lubman. D.M., 1987, *Anal. Chem.* 59:31A.
Lubman, D.M., 1993, Two-step methods for separated volatilisation and laser-induced multiphoton ionisation of small biological molecules in supersonic jets, *in*: "Laser Ionisation Mass Analysis,"

Akos, V., Renaat, G., and Adams, F., eds., Chemical Analysis Series, Vol. 24, John Wiley and Sons Inc., New York, pp. 321-368.

Maechling, C.R., Clemett, S.J., Zare, R.N., and Buseck, P.R., 1993, *in*: "Proc. 41st ASMS Conf. Mass Spectrometry and Allied Topics," Washington DC, 428.

Mowat, I.A., and Donovan, R.J., 1995, *Rapid Commun. Mass Spectrom.* 9:82.

Nordhoff, E., Ingendoh, A., Cramer, R., Overberg, A., Stahl, B., Karas, M., Hillenkamp, F., and Crain, P.F., 1992, *Rapid Commun. Mass Spectrom.* 6:771.

Posthumus, M.A., Kistemaker, P.G., Meuzelaar, M.C., and Ten Noever de Brauw, M.C., 1978, *Anal. Chem.,* 50:985.

Schlag, E.W., Grotemeyer, J., and Levine, R.D., 1992, *Chem. Phys. Lett.* 190:521.

Strupat, K., Karas, M., and Hillenkamp, F., 1991, *Int. J. Mass Spectrom. Ion Processes* 111:89.

Sundqvist, Bo.U.R., 1992, *Int. J. Mass Spectrom. Ion Processes* 118/119:265.

Tanaka, K., Waki, H., Ido,Y., Akita,S., Yoshida,Y., and Yoshida, T., 1988, *Rapid Commun. Mass Spectrom.* 2:151.

Tembreull, R., and Lubman, D.M., 1986, *Anal. Chem.* 58:1299.

van Bremen, R.B., Snow, M., and Cotter, R.J., 1983, *Int. J. Mass Spectrom. Ion Phys.* 49:35.

van Veelen, P.A., Tjaden, U.R., van der Greef, J., Ingendoh, A., and Hillenkamp, F., 1993, *Chromatogr.* 647:367.

Vertes, A., and Gijbels, R., 1993, Laser-induced thermal desorption and matrix-assisted methods, *in*: "Laser Ionisation Mass Analysis," Akos, V., Renaat, G., and Adams, F., eds., Chemical Analysis Series, Vol. 24, John Wiley and Sons Inc., New York, pp. 127-175.

Weyssenhoff, H.V., Selzle, H.L., and Schlag, E.W., 1985, *Z. Naturschforsch.*, 40a:674.

Wu, K.J., Steding, A., and Becker, C.H., 1993, *Rapid Commun. Mass Spectrom.* 7:142.

Zare, R.N., and Levine, R.D., 1987, *Chem. Phys. Lett.* 136:593.

Zenobi, R., 1994, *Chimia* 48:64.

MULTIPHOTON IONIZATION AND LASER MASS SPECTROMETRY

Kenneth W.D. Ledingham

Department of Physics and Astronomy
University of Glasgow
Glasgow G12 8QQ
Scotland

INTRODUCTION

Laser mass spectrometry is an ultra-sensitive analytical technique with a great many applications (Lubman, 1990; Vertes et al., 1993). An explanation of the many acronyms that abound in laser mass spectrometry will be given but with particular reference to resonance ionization mass spectrometry (RIMS) or resonance enhanced multiphoton ionization (REMPI), or more generally laser post-ionization of neutral atoms/molecules. This technique will be distinguished from other important similar technologies e.g. secondary ionization mass spectrometry (SIMS) and laser microprobe mass spectrometry (e.g. LIMA/LAMMS).

Resonance Ionization Mass Spectrometry (RIMS) is a particular case of the more general technique of laser mass spectrometry and since there is a great deal of laser spectroscopy involved, it will be given prominence in this paper. RIMS is a relatively new analytical technique which has only been developed in the last ten years or so. The physics on which it is based is, however, much older. Göppert-Mayer (1931) derived the basic expression for the two-photon transition rate (the simplest RIMS process) from second order perturbation theory as long ago as 1931. A sensitive laser spectroscopic technique was proposed by Ambartzumian and Letokhov (1972), more than twenty years ago, but it is only in the last decade that the pioneering work of Letokhov (1987) and Hurst and Payne (1988) has come to fruition.

In RIMS there are many arrangements of lasers and mass spectrometers. The lasers are used for two purposes. Firstly, they gasify or atomise (by desorption/ablation) solid samples and secondly, they ionize resonantly or non-resonantly the emitted neutral atoms or molecules. The mass spectrometers can be of magnetic, quadrupole or much more commonly time of flight types.

Acronyms abound in the field of laser mass spectrometry and with the aid of Fig.1 a number of these will be explained. When an ion beam hits a target and sputters off ions the procedure is called secondary ionisation mass spectrometry (SIMS). SIMS is one of the most important trace analytical techniques in use to today, especially for semiconductor and

An Introduction to Laser Spectroscopy
Edited by David L. Andrews and Andrey A. Demidov, Plenum Press, New York, 1995

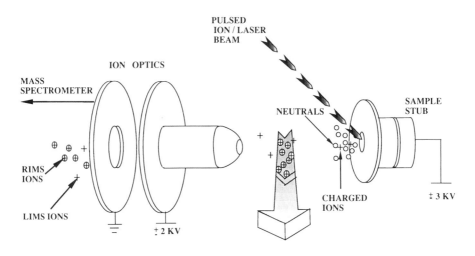

RESONANT OR NON RESONANT
LASER BEAMS POST IONISING NEUTRALS

Figure 1. Ions and neutral atoms are emitted from a surface after bombardment with an ion or laser beam. The neutral atoms are post-ionized with a second laser beam as shown.

polymer analyses. If a pulsed laser beam strikes the target and desorbs/ablates positive or negative ions the process is called laser-microprobe mass analysis (LIMA or LAMMS). In the sputtering or ablation process, the yield of ions is several orders of magnitude lower than the production of neutral atoms or molecules. Where the neutral plume is post-ionized by a second laser system the process has been called Surface Analysis by Laser Ionization (SALI, Becker and Gillen, 1984) if the laser wavelength is off-resonance and in the UV, and Resonance Ionisation Mass Spectrometry (RIMS) or Laser Ablation Resonant Ionization Spectrometry (LARIS, Arlinghaus et al., 1991) if the lasers are resonant. If the plume consists of molecules rather than atoms the process is usually called Resonance Enhanced Multi-Photon Ionization (REMPI).

The reasons that led to the development of RIMS, or more generally post-ionization of neutrals, are firstly that the sensitivity is likely to be increased over LIMA and SIMS since the number of neutral atoms or molecules is between two and three orders greater than the number of ions and, secondly, that the matrix problems associated with RIMS are greatly reduced. Matrix problems (lack of quantification) in SIMS can be at the level of some orders of magnitude while for RIMS it has been shown to be only about a factor of two.

MULTIPHOTON IONIZATION

Atoms of each element have a unique set of excited states which may be reached from the ground state by the absorption of one or more photons at the correct frequency providing certain optical selection rules are satisfied. An excited atom would normally decay back to the ground state with a characteristic time of about 10 ns. However before decaying, the atom can absorb another photon which may take the atom to a higher excited state or cause ionization.

The cross section for resonant absorption of one photon between bound levels is typically of the order of 10^{-16} m^2 and a laser fluence of about 1 mJ m^{-2} is sufficient to cause saturation excitation of the atoms in the laser beam. The cross sections for photoionization are normally 10^{-6} times smaller and laser fluences of about 1 kJ m^{-2} are required for saturation. Saturation here signifies excitation or ionization of every atom or molecule in the laser beam. This two-photon ionization process is shown in Fig.2a. Most modern tunable lasers require moderate focusing to reach these values. Two other ionization procedures have been employed to alleviate this problem of high fluences (Bekov and Letokhov, 1983). These are shown in Fig.2 b,c. In (b) the atom is excited to a Rydberg state and is finally ionized by a pulsed electric field with high efficiency. The other method (c) is to ionize the atom via autoionization states. The rate limiting step in (b), (c) has a cross section some two orders of magnitude larger than in (a) and requires much lower fluences to reach saturation.

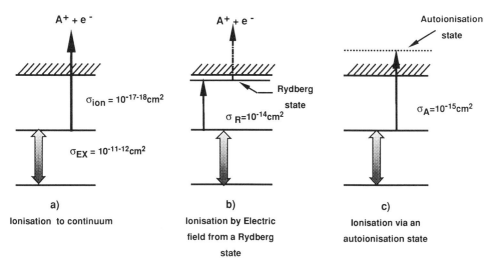

Figure 2. (a) An electron in its ground state absorbs a photon and is raised to an excited state. Ionization by absorption of a second photon has a small cross section and is the rate limiting step: (b) The atom is excited to a Rydberg state and is finally ionized by a pulsed electric field with high efficiency: (c) Ionization via an autoionization state with a large cross section is the preferred route.

The laser linewidth typically used in RIMS measurements is between 0.1 and 1.0 cm^{-1} and hence all the isotopes of an element are ionized simultaneously. Separation of the various isotopes is achieved in the mass spectrometer.

All atomic transitions between the ground state and some excited states, with the exception of He and Ne, may be resonantly excited with commercially available tunable dye lasers. Hurst and Payne (1988) have proposed five basic ionization schemes according to the relative energy positions of the intermediate states to the continuum. These are shown in Fig.3. In the first scheme a level exists at an energy that is more than half the value required to ionize the atom. Hence the atom can be ionized by two photons from the same dye laser, with the first photon being resonant. In the second scheme the output of the tunable dye laser must be frequency doubled to resonantly excite the atom, which is then ionised by a photon from the more intense fundamental beam. Fig.4 shows the whole Periodic Table with one of the schemes being ascribed for each element.

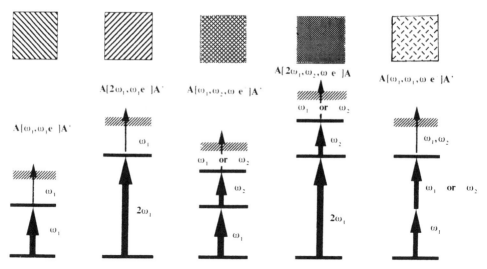

Figure 3. Five laser schemes which can ionize every element in the Periodic Table except helium and neon.

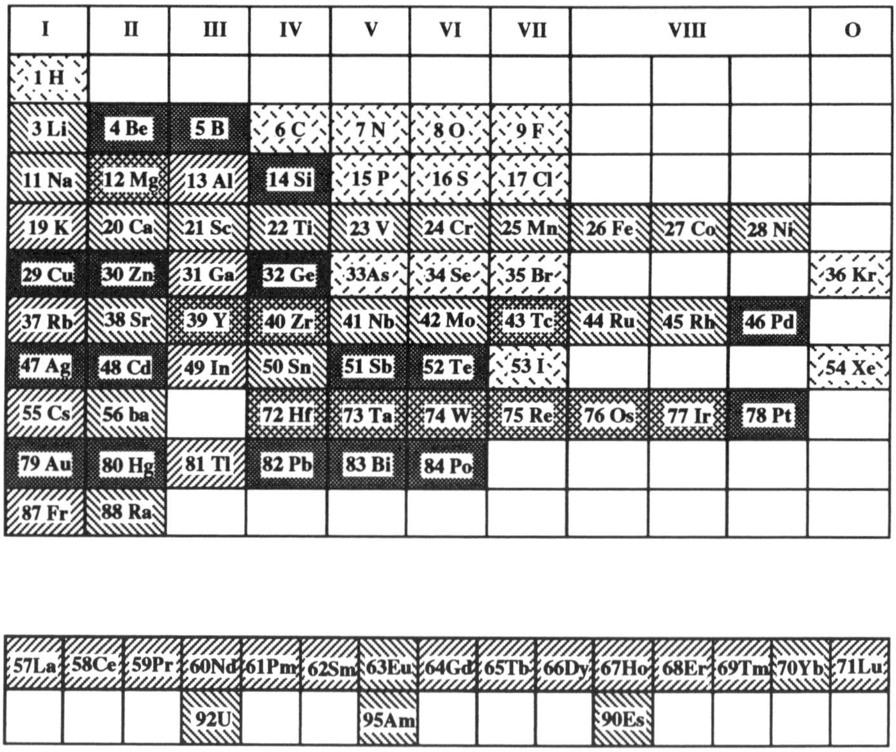

I	II	III	IV	V	VI	VII	VIII			O
1 H										
3 Li	4 Be	5 B	6 C	7 N	8 O	9 F				
11 Na	12 Mg	13 Al	14 Si	15 P	16 S	17 Cl				
19 K	20 Ca	21 Sc	22 Ti	23 V	24 Cr	25 Mn	26 Fe	27 Co	28 Ni	
29 Cu	30 Zn	31 Ga	32 Ge	33As	34 Se	35 Br				36 Kr
37 Rb	38 Sr	39 Y	40 Zr	41 Nb	42 Mo	43 Tc	44 Ru	45 Rh	46 Pd	
47 Ag	48 Cd	49 In	50 Sn	51 Sb	52 Te	53 I				54 Xe
55 Cs	56 ba		72 Hf	73 Ta	74 W	75 Re	76 Os	77 Ir	78 Pt	
79 Au	80 Hg	81 Tl	82 Pb	83 Bi	84 Po					
87 Fr	88 Ra									

57La	58Ce	59Pr	60Nd	61Pm	62Sm	63Eu	64Gd	65Tb	66Dy	67Ho	68Er	69Tm	70Yb	71Lu
			92U			95Am				90Es				

Figure 4. The Periodic Table with an appropriate scheme for each element (Hurst and Payne, 1988).

It can be seen from Fig.4 that many elements can be ionized using a single dye laser with a frequency-doubling capacity. For complete elemental coverage, one requires a large pump laser, an excimer or Nd:YAG laser and two tunable lasers, one which is frequency doubled. In fact with a single dye, 39 elements can be ionised, in principle enabling computer controlled element changes in a matter of seconds (Thonnard et al., 1989).

RIMS is an acronym primarily associated with the detection of atoms which usually have a simple set of energy levels. The same techniques can be used with molecules, but molecular structure is considerably more complex, since each electronic level has an associated set of vibrational and rotational levels. For molecules the cross sections for bound-bound and the bound continuum transitions have about the same order of magnitude at about $10^{-22}\,m^2$ (Boesl et al., 1981).

Fig. 5 shows several different molecular ionization schemes. It should be remembered that for each vibrational level shown in Fig.5, there is a manifold of rotational levels that often generates a continuous run of available excitational levels up to the ionization level. Because of this complexity the absorption features of some molecules at room temperature are in general broad and structureless, with a consequent loss in selectivity. To improve selectivity these molecules can be cooled down by supersonic or jet cooling to produce a few sharp peaks. This cooling is carried out by seeding the molecules into a light carrier gas (e.g. He or Ar at a few atmospheres pressure) and expanding into vacuum through a narrow orifice.

It is worthwhile to consider in some detail the rate equations involved in a simple two photon resonant ionization process (Fig.2a) since this gives an insight into the requirements of the resonant and non-resonant lasers used in laser post-ionization experiments (Hurst et al., 1979; Singhal et al., 1989).

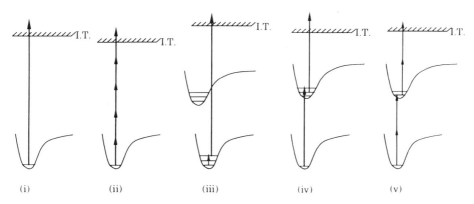

Figure 5. Several different ionization schemes for molecules. (1) Single photon ionization. (2) Non-resonant multiphoton ionization (MPI) - 5 photons are involved. (3) Resonant two photon ionization via ground state rovibrational levels. (4) Resonant two photon ionization via excited state rovibrational levels. (5) Resonance enhanced multiphoton ionization (e.g. two photon excitation, two photon ionization).

Let $N_0(t)$ be the total, time-dependent number of ground state atoms in the laser beam and $N_1(t)$ the corresponding number in the 1st excited state, $N_I(t)$ the number of ionized atoms. Also let σ_A be the stimulated absorption cross section at the ground state to state 1 frequency, σ_I the ionization cross section from state 1 to the continuum and τ the mean lifetime for spontaneous emission from state 1. In this very simple derivation the laser beam is considered to be monochromatic and of square pulse length T, with a flux (number of photons per unit area per unit time) Φ and a fluence (number of photons per unit area) ΦT. Thus we have

$$\frac{dN_0}{dt} = -N_0\sigma_A\Phi + \frac{N_1}{\tau} + N_1\sigma_A\Phi \tag{1}$$

$$\frac{dN_1}{dt} = N_0\sigma_A\Phi - \frac{N_1}{\tau} - N_1\sigma_A\Phi - N_1\sigma_I\Phi \tag{2}$$

$$\frac{dN_1}{dt} = N_1\sigma_I\Phi . \tag{3}$$

By adding equations (1) and (2) and also differentiating (2) with respect to t we get

$$\frac{d^2N_1}{dt} + [(2\sigma_A + \sigma_I)\Phi + \frac{1}{\tau}]\frac{dN_1}{dt} + \sigma_A\sigma_I\Phi^2N_1 = 0 \tag{4}$$

and the solution of this second order differential equation is:-

$$N_I = \frac{N_0\sigma_I\sigma_A\Phi^2}{\lambda_2 - \lambda_1}[\frac{1}{\lambda_2}\{\exp(-\lambda_2 T) - 1\} - \frac{1}{\lambda_1}\{\exp(-\lambda_1 T) - 1\}]$$

with
$$\lambda_2 = b + \sqrt{b^2 - \omega^2} ,$$
$$\lambda_1 = b - \sqrt{b^2 - \omega^2} ,$$
$$2b = (2\sigma_A + \sigma_i)\Phi + \frac{1}{\tau} ,$$
$$\omega^2 = \sigma_A\sigma_I\Phi^2$$

A typical graph of N_I against laser flux is shown in Fig.6 and it can be seen that at low fluxes the gradient is 2, decreasing to 1, and then reflecting saturatation i.e. where all atoms in the laser beam volume are ionized, and $N_I = N_0$. The saturation conditions are:-

$$\sigma_A\Phi \gg \frac{1}{\tau} \qquad \sigma_I\phi \gg 1 \quad .$$

For bound-bound transitions saturation occurs at about 1 mJ m^{-2} and for bound-continuum transitions about 1 kJ m^{-2}.

TYPICAL EXPERIMENTAL ARRANGEMENTS FOR LASER MASS SPECTROMETRY

Typical experimental arrangements for laser microprobe mass analysis and post-ionization techniques will be described in this section. Only a brief description of the laser microprobe will be given here.

The analysis of solids using a time of flight mass spectrometer to detect ions generated by a laser is not a new idea. The initial laboratory experiments were carried out by Fenner and Daly (1966) and the history of the laser microprobe (LAMMS or LIMA depending on the manufacturer) has been reviewed and assessed by Clarke (1989) and Ledingham (1992). A diagram of one of the most popular laser microprobes is shown in Fig.7.

In the LIMA instrument, ablation and ionization are normally carried out using a focused Nd:YAG laser operating at its fundamental wavelength (1.064 nm) or one of its harmonics, doubled (532 nm), tripled (355 nm) or quadrupled (266 nm) although for good

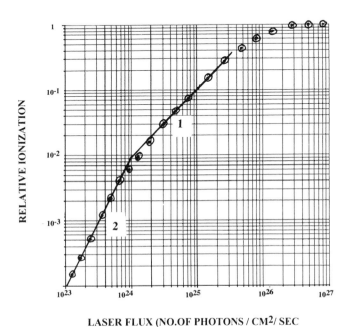

LASER FLUX (NO.OF PHOTONS / CM^2/ SEC

Figure 6. Typical graph of number of ions as a function of laser flux (equation 4). Saturation approaches above 10^{30} photons $m^{-2} s^{-1}$. The numerals 2 and 1 refer to the gradients of the graph in these regions.

ionization efficiency 266 nm is preferred. The laser flux can be varied between 10^{10-15} W m^{-2} with a pulse length typically of 5 ns and a spot size between 0.5 – 5 μm when the laser is operated in TEM$_{oo}$ mode. The typical sample volume consumed in this process ranges from 0.01 μm^3 at desorption threshold to about 5 μm^3 at maximum useful flux. By biasing the sample probe either positively or negatively with respect to the grounded extraction electrode, positive or negative ions are extracted and transmitted through a reflectron time of flight mass spectrometer with a resolving power of a few thousand. The elemental detection limits are between 0.1 and 100 ppm and complete elemental and isotopic detection between hydrogen and uranium is possible.

Schematic diagram of LIMA

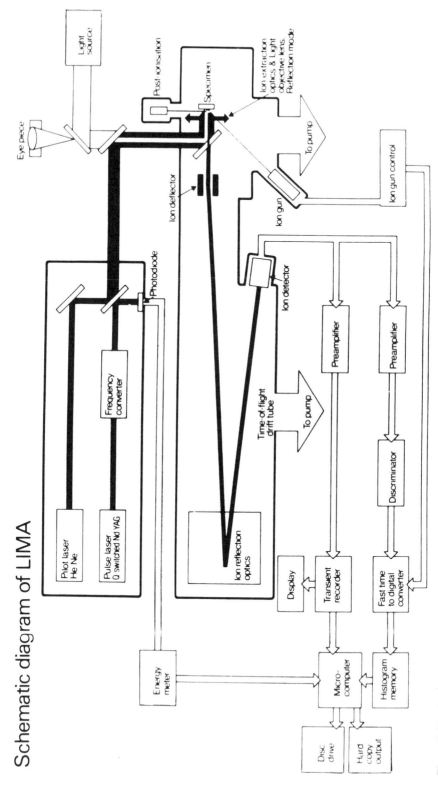

Figure 7. The CMS LIMA instrument. (Reproduced by permission of CMS Ltd). Usually the laser induced ions alone are detected in this instrument although the advanced instruments have a post-ionization facility.

A)

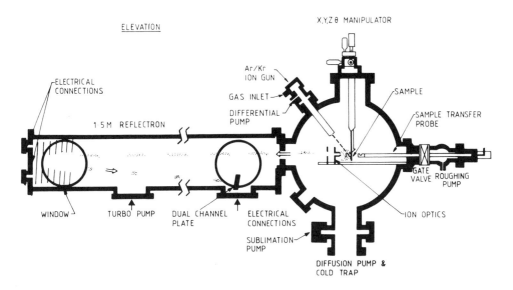

B)

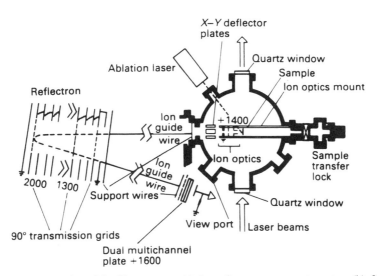

Figure 8. (a) Elevation view of the Glasgow post-ablation reflectron mass spectrometer. (b) View of the spectrometer showing the electrostatic ion reflector in greater detail. An initial spread of ion energies is compensated by the reflectron since the high energy ions penetrate deeper into the reflecting field and hence spend a longer time than the lower energy ions. This improves the resolution and the guide wire increases the transmission. A recent modification permits controlled ion fragmentation by a further laser beam at the turn-round point.

There are many arrangements for post-ionization mass spectrometry (RIMS) analysis using different types of mass spectrometers – magnetic sector, quadrupole and time of flight of both linear and reflectron types. There are also many types of lasers used, both of cw and pulsed designs. By far the most common configuration is a pulsed laser arrangement coupled to a time of flight mass spectrometer.

The Glasgow University design (Towrie et al., 1990) described here is a typical RIMS system which is now available commercially. Fig.8 shows the system in detail. The spherical sample chamber is 30 cm in diameter with as many ports as possible facing the centre of the chamber, the point at which the sample stub is held. The sample is mounted on an xyz manipulator and can be inserted and withdrawn from the sample chamber using a rapid transfer probe. Fast sample exchanges (5 mins) can be made without disruption of the main chamber pressure. The sample chamber is pumped by an oil diffusion pump fitted with a cold trap and a titanium sublimation pump, having a base pumping speed of over 800 ls^{-1} and capable of maintaining a base pressure of less than 10^{-9} torr in the chamber. The ion extract optics and the electrostatic reflector are shown in Fig.8(b). The sample is maintained at a voltage of about 2000 V with the first extraction electrode at 1400 V. The reflectron time of flight (TOF) system has an overall drift length of 3 m.

The principal factor that limits the resolution of a conventional TOF is the spread of the initial ion/neutral energies in the ablation process. This spread of ion energies can be compensated using a reflectron time of flight mass spectrometer in which the high energy ions penetrate deeper into an electrostatic ion reflector and hence experience a longer flight time than the ions of lower energy. The FWHM of the present system is about 1000 for ions of about 40 amu. A thin wire, 0.005 cm in diameter, follows the ion path through the flight tube, providing an electrostatic guide for the ions, increasing the transmission of the mass spectrometer. The wire operates at about –10 V.

Ablation is carried out using a Quantel Nd:YAG laser operated either at 1064 nm or one of the harmonics, although normally 532 nm is the preferred wavelength. If sputtering is used the system is fitted with a Kratos Penning ion gun. Both inert gases and oxygen ions can be used with energies up to 15 keV and with beam currents up to 5 μA.

The post-ablation ionization laser system consists of a Spectron Nd:YAG laser powering two dye lasers, one of which can be frequency doubled. The lasers operate at repetition rates of 10 Hz and with pulse duration of about 10 ns. The tunable laser pulse energies depend on the dyes used but are normally between 100 μJ and 2 mJ. The time between the ablation/sputter system can be varied between 0.1 and 10 μs, although typically this is 1 μs. The ionizing laser is introduced into the sample chamber parallel and as close as possible to the sample stub to maximise the overlap with the ablation plume.

The ions after passing down the TOF are detected in a dual channel plate detector and passed to a data acquisition system which measures and stores mass spectra and laser pulse energies on a pulse to pulse basis. A Lecroy 2261 transient recorder or 9310 digital oscilloscope coupled to a COMPAQ 386/25 forms the basis of the system. Ion signals from the multichannel plate detector are digitised by the transient recorder providing 640 time channels (11 bit resolution) each of 10-100 ns width. Both wavelength dependent spectra and mass spectra at specific wavelengths can be taken with the data acquisition system.

The RIMS instrument described above is being routinely used for trace analysis and for studies of the characteristics of laser ablation from solid samples. Fig.9(a) shows a RIMS signal for gold in copper at 10 ppm level, accumulated over 10^4 shots. Fig.9(b) indicates how linear the technique is over a wide range of concentrations from majors to minors for gold in copper (McCombes et al., 1991). RIMS trace detection down to parts per billion is now common in semiconductors, and indeed sensitivities down to parts in 10^{12} have been reported for specific elements (Bekov et al., 1985). Atom Sciences Inc. can now make noble gas detection measurements to levels below 100 atoms in five minutes which makes solar neutrino, small meteorite and ground water dating experiments feasible (Thonnard et

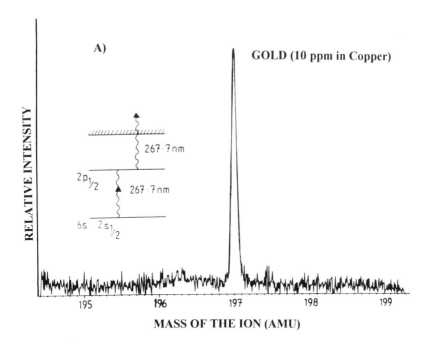

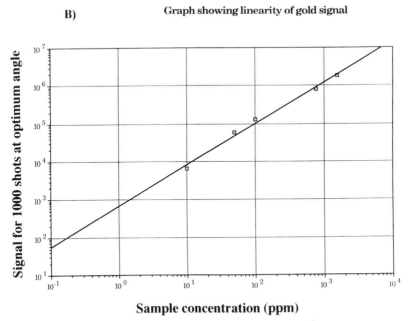

Figure 9. (a) RIMS signal for gold in copper at 10 ppm, accumulated over 10^4 shots. A two photon scheme at 267.7 nm is used. (b) RIMS signal versus gold in copper concentration illustrating the linearity of the process.

al., 1992). Attogram detection limits and isotopic selectivity greater than 10^9 have been demonstrated for rare radionuclides using cw laser RIMS (Bushaw, 1992).

SIMS is a very mature analytical technique that is widely used for characterising materials. It has, however, two limitations which are not shared by RIMS. First, there is the very strong matrix effect, the often unknown relationship between the chemical composition of a solid and the secondary ion yield of the analyte in the solid. Secondly, another difficulty with SIMS is its inability to distinguish isobaric interferences. SIMS cannot easily distinguish isobars unless the resolution of the mass spectrometer is extremely high. RIMS on the other hand easily distinguishes isobars, since the ionizing laser is tuned to transitions in a specific element.

Fig.10 shows a sputter initiated RIMS analysis on a series of samples with silicon in different matrices e.g steel, niobium and tungsten (Parks et al., 1986). This is a remarkable graph which could not be obtained using SIMS. The graph is reasonably linear and the lack of matrix dependence is attributed to separating in time and space the sputtering and the ionisation processes. In the SIMS procedure these processes are coupled.

A number of laboratories have demonstrated the extreme sensitivity and selectivity of RIMS, but less effort has been devoted to the questions of accuracy and precision which are normally as important to analysts. RIMS of sputtered atoms (SIRIS) has the potential of being a quantitative ultra-trace technique. The reason for this is that the sputtering of clean metals by noble gas ions produces a plume dominated by ground state neutral atoms. In addition there is large database of sputtering rates. If the sputtering rate, the transmission of the mass spectrometer and the degree of laser ionization are all known, then it is possible to draw quantitative conclusions regarding impurity concentrations.

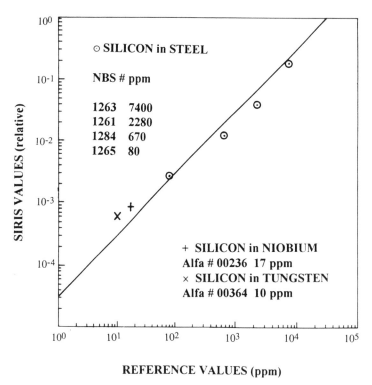

Figure 10. Comparison of sputter initiated resonance ionisation spectrometry (SIRIS) measurements using certified values of silicon in different matrices (Parks et al. 1986). (Reproduced by permission of the authors and the Institute of Physics.)

Using an ion sputtering plus RIMS instrument, Argonne National Laboratory have recently analysed a number of well characterised metallic samples containing iron, copper, molybdenum, indium and lead. In addition Si samples contaminated with Cr, Cu, Ni, Fe and also with implanted Mo were analysed by RIMS and compared with TXRF and known implantation rates. The significant finding is that trace impurity measurements down to ppb levels in metal matrices can be made quantitative by employing pure metal targets as calibration standards. This discovery substantially reduces the effort required for quantitative analysis, since a single standard can be used for determining concentrations spanning 9 orders of magnitude. A comparison of the different techniques for the contaminated silicon produced agreement to within a factor of three, down to parts per billion sensitivity (Calaway et al., 1992). It would appear that RIMS has the potential to become an ultra-trace quantitative analytical technique.

SOME IMPORTANT APPLICATIONS FOR LASER MASS SPECTROMETRY

Matrix Assisted Laser Desorption (MALDI) for Heavy Biomolecules

With the fast development of peptide synthesis and genetic engineering (e.g. DNA sequencing and the Human Genome problem) in particular, the need arose for a rapid, accurate and sensitive detection technique for determining the molecular weight of proteins and polypeptides.

Laser desorption of large biomolecules without fragmentation was a technique, of considerable promise in the early eighties, which scarcely lived up to expectations. However in 1987, Tanaka et al. and Karas et al. developed a new variation of laser desorption, which they called matrix assisted laser desorption/ionization (MALDI), to study large inorganic molecules. This technique, one of the real growth areas in laser mass spectrometry, is now capable of desorbing intact biomolecules of up to 500 kDa (Dominic Chan et al, 1992) with excellent sensitivity and without lengthy sample preparation. The resulting ions are then analysed in a mass spectrometer of linear or more commonly reflectron type.

The technique involves focusing a pulsed laser, usually UV and often a nitrogen laser (337 nm), on to crystals consisting of a suitable host material (the matrix) doped with the sample of interest (the analyte) with a matrix to analyte concentration typically of about 1000/1. As the technique now stands the small concentration of high mass analyte, usually exhibiting only a low absorption of the laser light per molecule, is imbedded in either a solid or liquid of low mass with high absorption of the laser light. In this way there is an efficient and controllable transfer of energy to the analyte molecules without the extensive fragmentation of heavy masses which would happen with laser desorption in the absence of the matrix.

Although the practical aspects of MALDI are known, the underlying physical principles involved in the formation of the analyte ions are uncertain. It is suspected that suitable matrices are photoexcited or ionized in the resonant absorption process and then transfer a proton to the analyte in the gas phase. This is still very much a hypothesis, although recent experiments in Glasgow suggest that considerable numbers of neutral hydrogen atoms are liberated in the ablation process (Scott et al., 1994).

The laser procedure and apparatus for carrying out a MALDI analysis is similar to the LIMA/LAMMS procedure. A crystal containing matrix and analyte is deposited on the sample stub and ablated at very low power levels of about 10^{10} W m^{-2}, usually using a focused nitrogen or quadrupled Nd:YAG laser. The procedure is essentially a very soft ablation process. The desorption/ionization process is such that both positive and negative ions are generated for many molecules (peptides and proteins) with similar signal intensities. One of the differences from the normal LIMA/LAMMS procedure is that since the gain of multichannel plates decreases with higher masses because of their low impact velocity, very high acceleration potentials or post-acceleration fields after the drift region (20 - 30 kV) are necessary to increase the sensitivity. Fig.11 shows laser ablation of substance P (a peptide of mass 1348 Da) with and without the assistance of a light mass matrix 2,5-dihydroxibenzoic acid (DHB). It can be seen that with matrix assistance, there is a large parent ion peak, but without the matrix there is hardly any parent ion. In Fig.12 a high mass spectrum is shown and improved signal to noise ratios are obtained by signal averaging (Hillenkamp, 1994).

The potential of MALDI in the bioanalytical field is considerable and already it is considered to be the state-of-the-art technique for the detection of such different chemical substances as polypeptides, proteins, oligonucleotides, polysaccharides and sulfonic acids. Masses up to 500 kDa can be measured with an accuracy of about 0.2% and even less from sample sizes as small as 100 fmol (Olaf-Bornsen et al., 1990).

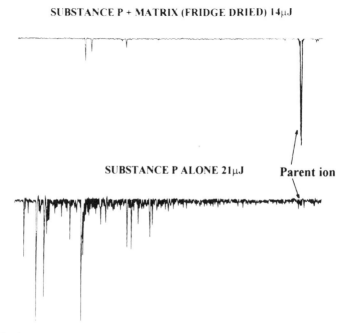

Figure 11. The importance of the matrix for parent mass ion production. Without the matrix almost no parent mass is visible although there is considerable fragmentation. The laser wavelength is tuned to the maximum absorption of the matrix.

Resonant Laser Ablation for Surface Analysis

As has been described, LIMA/LAMMS is a well established laser microprobe trace analytical technique in which a focused laser is directed at the surface of a sample. The ions created are analysed normally by a reflectron time of flight mass spectrometer to ppm sensitivities or lower. Its speed and versatility make it a very useful instrument. It has also been described (Ledingham, 1992) that greater sensitivity can be obtained by post-ablation ionization of the neutral atoms or molecules in the ablated plume using a two laser system, one to ablate and one to ionize (either resonantly or non-resonantly).

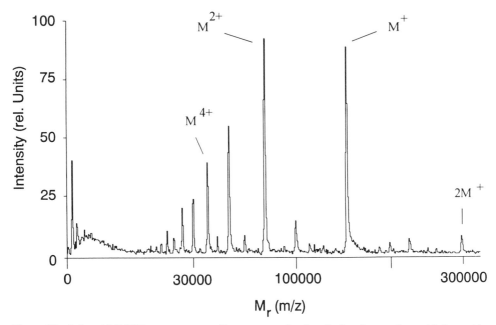

Figure 12. Infrared-MALDI mass spectrum of human monoclonal antibody. Averaged over 15 shots with a measured molecular weight of 149500 Da and a mass resolution of 150. An Er-YAG laser was used with wavelength 2.94 µm and pulse width 150 ns. (Reproduced by permission of F Hillenkamp).

In a series of experiments carried out at Glasgow University, it has been shown that both ablation and resonant ionization may be performed with a single laser pulse to provide ionization enhancements of several hundred for a number of different atoms. This procedure has been called resonant laser ablation (RLA). In the scheme shown in Fig.13, for example, a pulsed laser beam from a tunable dye laser was directed usually at grazing incidence (~5°) to a series of samples -GaAs, AlGaAs, and NIST steels with traces of aluminium. The dye laser output was frequency doubled and both the fundamental and the doubled beams were moderately focused with a 30 cm lens to a diameter of 0.5 mm, resulting in an irradiation area on the sample surface of about 2.5×10^{-3} cm^2. The ions created were analysed in time-of-flight mass spectrometers of either linear or reflectron type. The ion signal for Al from AlGaAs as a function of the ablation laser wavelength is shown in Fig.14. The resonance effects are clearly seen with a signal enhancement of greater than two orders of magnitude. Careful calibration of the laser wavelength indicated that the resonances correspond to known atomic transitions and are sharp, indicating that the neutral atoms were first desorbed from the surface of the sample, resonantly excited and then subsequently ionized in the gas phase by the same laser pulse.

This is therefore a two step process similar to post-ablation ionization techniques (LARIS) and completely different from LIMA. One of its advantages is that it requires a simpler experimental arrangement than LARIS. Recent work has suggested that the width of the peaks, increasing with both increasing laser fluence and angle of incidence, is at least in part caused by atomic collisions within the plume (Wang et al., 1992 and all the references therein).

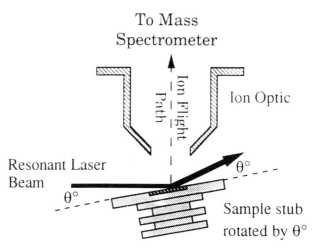

Figure 13. The geometric arrangement for RLA measurements with the laser beam at grazing incidence (θ) to the sample stub.

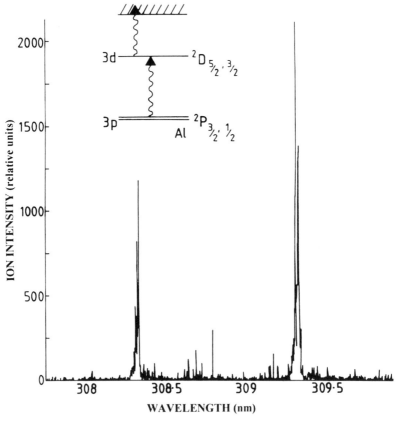

Figure 14. The ion signal for Al as a function of the ablation laser wavelength. The enhanced yield at wavelengths corresponding to the excitation of the 3d level from the ground state doublet is clearly observed.

a) NIST SRM 1263A (On Resonance)

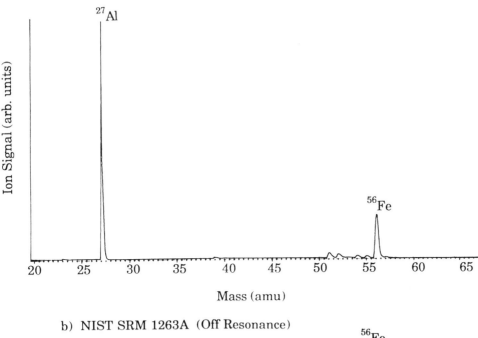

b) NIST SRM 1263A (Off Resonance)

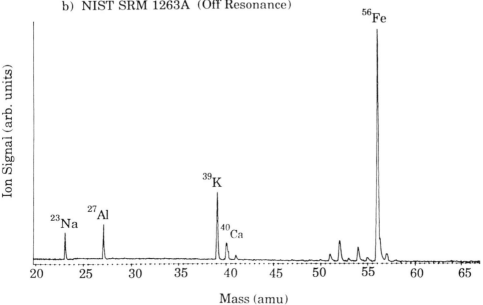

Figure 15. (a) Spectrum of NIST SRM 1263A with the laser tuned to the Al resonance at 308.3 nm for 500 shots. (b) Off-resonance spectrum at 307.5 nm: scale **x** 10 compared to (a).

All of the initial work on RLA was carried out on majors. A steel sample (NIST SRM 1263A) containing 5000 ppm Al was chosen to demonstrate that the technique was also sensitive to minor levels. Mass spectra were obtained on and off resonance and are shown in Fig. 15. It can be seen that the yield of Al ions is highly enhanced at resonance and a sensitivity of a few ppm has been estimated from the signal to noise ratio from this data. A second steel sample (NIST SRM 1261A) containing 500 ppm Al was also analysed and produced an aluminium signal five times smaller than the 1263A sample. The agreement with the nominal ratio (10:1) is considered to be reasonable at this stage of the quantitative development of RLA (Borthwick et al., 1992).

RLA has been demonstrated by the Glasgow group for calcium, aluminium, gallium, iron silicon and manganese, from a variety of surfaces, shiny to rough, and from metals, semiconductors and non-conductors. A recent review paper of RLA by Eiden et al. (1994) has shown that 12 elements can be analysed using a very inexpensive nitrogen pumped dye laser system, with 100 shots and 20 µJ per shot, with sensitivities less than 100 ppb. They have also shown that since the ablation process is non-thermal, highly excited species are easily detected allowing spectroscopic studies of transitions that originate in states which are inaccessible (spin and parity forbidden) from the ground state.

Trace Detection of Radio Toxic Isotopes by RIMS

RIMS has become an important technique for ultra-trace environmental analysis, and in particular it is well suited for the detection of long-lived radio toxic isotopes. The applications of RIMS to the trace detection of radioactivity has been reviewed recently (Wendt et al., 1994 and Payne et al., 1994) and in this review only the application to the

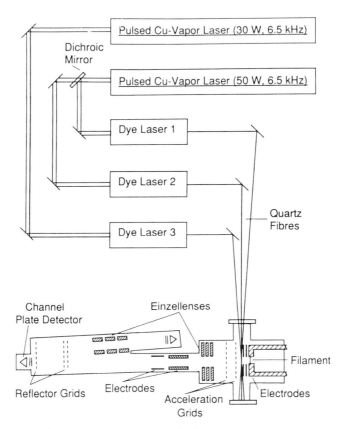

Figure 16. Experimental arrangement for trace analysis of plutonium and other actinides evaporated from a filament. The atoms are resonantly excited and ionized using three dye lasers pumped by two copper vapour lasers. The ions are mass selectively detected with a reflectron TOF.

detection of plutonium will be described, although the Mainz group have also developed sophisticated techniques for the detection of technetium and strontium isotopes.

The experimental arrangement for the trace detection of the actinides including plutonium is shown in Fig. 16. The samples are thermally desorbed from a filament in the source region of a time of flight mass spectrometer. The neutral atoms are then ionized with three tunable dye lasers, pumped by two copper vapour lasers at repetition rates of 6.5 kHz. The dye laser beams are coupled into the ionization region of the TOF either with prisms or by optical fibres. The ions produced by resonant laser excitation and ionization in the interaction region are accelerated to an energy of 3 keV, pass down a drift length of 2 m and then are detected in a channel-plate detector. It was found that the best excitation scheme was $\lambda_1 = 586.49$ nm, $\lambda_2 = 665.57$ nm with the final step via an autoionizing state at $\lambda_3 = 577.28$ nm (Urban et al., 1992). The main losses limiting the overall efficiency are:- the

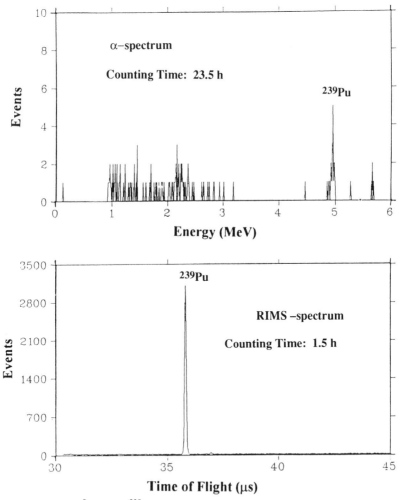

Figure 17. Detection of 10^9 atoms of ^{239}Pu. The upper graph is a conventional α-spectrum obtained in 23.5 hrs counting time and the lower is a RIMS spectrum of the same sample with a counting time of 1.5 hrs (Wendt et al., 1994). (Reproduced by permission of the authors.)

217

temporal overlap between the continuous atomization source and the pulsed lasers and the spatial overlap. A final typical efficiency for the detection of plutonium is 1×10^{-5}. Considering the background, this yields a detection limit of 2×10^{6} with a one hour detection time. For ^{239}Pu, this is an improvement by more than two orders of magnitude over conventional spectroscopy (4×10^{8} atoms in a typical measuring time of over 16 hours). Fig. 17 shows a comparison for the detection of 10^{9} atoms of ^{239}Pu by α-spectroscopy (top) and by RIMS (bottom).

In addition to rapid and sensitive analysis, the RIMS method permits the determination of isotopic abundances with an accuracy of only a few percent, which is not possible by direct radiometric methods.

Depth Profiling of Layered Semiconductors using RIMS

Depth-resolved information about dopant and impurity distributions in semiconductors like Si and GaAs is of great importance in the electronics industry. Techniques such as SIMS are very mature and are used routinely to characterise such materials. However for quantitative results, SIMS requires calibration with standards for every element in each matrix encountered. Alternatively for RIMS, assuming that the photoionization is saturated, the signals have been demonstrated to be almost element independent. It has been shown that equal amounts of impurities such as Be, Al and Co in Si give equivalent signals. This is what is meant by matrix independent. Fig.18 demonstrates this elemental independence (Downy and Emerson, 1991).

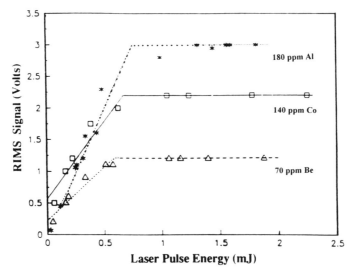

Figure 18. Demonstration of RIMS saturation of Al, Be and Co implanted into Si. Above 1mJ laser pulse energy, each element responds similarly and the signals are proportional to concentration (Downey and Emerson, 1991). (Reproduced by permission of the authors and John Wiley & Sons).

Some of the most difficult materials to characterise are compound semiconductors such as GaAs, AlAs, AlGaAs, InGaAs, and InAlAs, because devices use very thin layers of these materials. Fig. 19 shows a comparison between a SIMS and RIMS depth profile to detect beryllium in a layered structure of GaAs and AlAs. Alternative layers e.g. GaAs and AlAs, 100 nm thick, were grown on a GaAs substrate at 400°C and doped with Be to nominal uniformity at approximately 2×10^{19}cm^{-3}. It is known that under these growth conditions,

Be is ten times more soluble in GaAs than in AlAs. It can be seen that the RIMS profile represents the layered structure better than SIMS. Moreover, the Be spikes, known to accumulate at the interfaces, are of equal widths in the RIMS profile whereas in the SIMS data the widths depend on the ordering of the layers (Downy et al., 1990).

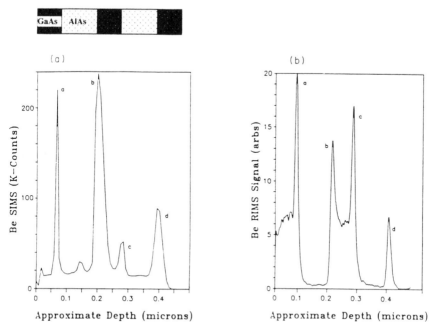

Figure 19. (a) SIMS (O_2^+ sputtering) profile of Be-doped GaAs/AlAs layers. (b) RIMS profile of Be in the same structure showing Be diffusion to the interfaces. The letters are used to identify corresponding points in the profile. The top layer and the layer between b and c are GaAs. (Downey et al., 1990). (Reproduced by permission of the authors and John Wiley & Sons.)

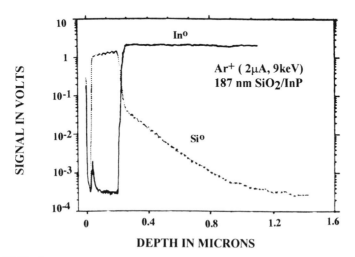

Figure 20. RIMS of sputtered neutral Si and In atoms in a depth profile through an Au/SiO$_2$/InP test sample. The signal has not been normalised: note the correct Si and In levels (Arlinghaus et al., 1990). (Reproduced by permission of the authors and the American Vacuum Society).

Fig. 20 shows a SIRIS Si/In depth profile through a Au-SiO-InP layered structure. The uncorrected SIRIS signal for Si and In have the approximately correct 1.0 to 1.5 ratio, indicating the essentially identical response to Si and In by the RIMS process (Arlinghaus et al., 1990).

It must be emphasised that RIMS is no way a competitor to SIMS. SIMS has been established for more than 25 years and is accepted as a very important technique for characterising materials - moreover the procedure at present is much simpler than RIMS. It is hoped that RIMS can be established as a complementary technique which can be used as a powerful aid for quantification of SIMS.

Ultra-trace Detection of Strategic and Environmentally Sensitive Molecules using REMPI

The most common RIMS scheme for analytical purposes using molecules is resonant two-photon ionization (R2PI) in which one photon excites the molecule to an electronic state and a second photon causes molecular ionization. Since molecules normally have

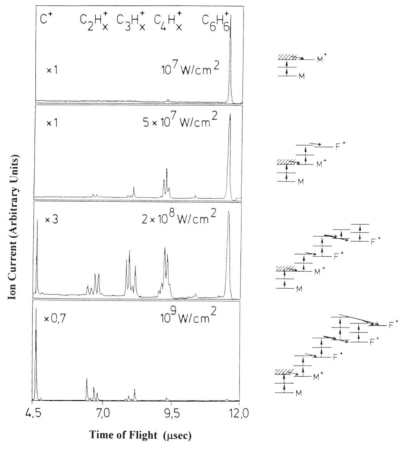

Figure 21. Time of flight mass spectra of the fragmentation of benzene due to multiphoton ionization processes at different laser intensities. On the right hand side is illustrated schematically the ladder switching for increasing laser intensities (Boesl, 1991). (Reproduced by permission of the author and the American Chemical Society).

ionization potentials between 7 and 13 eV, R2PI requires two UV photons. Tunable radiation of dye lasers with frequency doubling can provide UV radiation down to about 210 nm. Excimer lasers are powerful sources of UV and VUV, eg. XeCl (308 nm), KrF (248 nm), ArF (193 nm) and F_2 (155 nm) and are particularly useful for the photoionization step in R2PI.

In contrast to atoms, molecules in excited states can undergo different photophysical and photochemical transformations. Photodissociation can compete with photoexcitation and photoionization processes. For larger laser fluences, the fragmentation of the molecule is more extensive. Fig. 21 shows the fragmentation of benzene following an R2PI process, indicating at low fluence that only the parent peak is visible and at high fluence carbon ions are the most abundant fragments (Boesl et al., 1991). The fragmentation pattern gives important information regarding molecular structure. In addition the wavelength dependence of the parent or fragment ions can also be used as a fingerprint for trace detection of molecules.

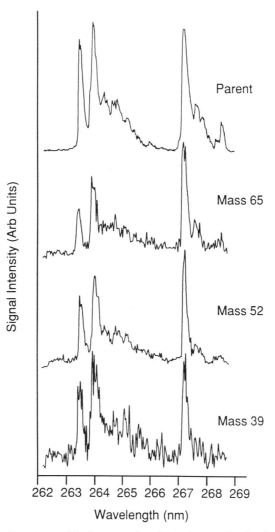

Figure 22. Wavelength dependence of the fragmentation of the toluene molecule. It should be noticed that the parent and a number of the prominent fragments have the same wavelength dependence.

Fig.22 gives a comparison of the wavelength dependence of the toluene parent ion with that of some of the more prominent fragment ions. From the great similarity of all the fragment spectra it can be concluded that fragmentation is initiated from the excited states of the parent (Marshall et al., 1991).

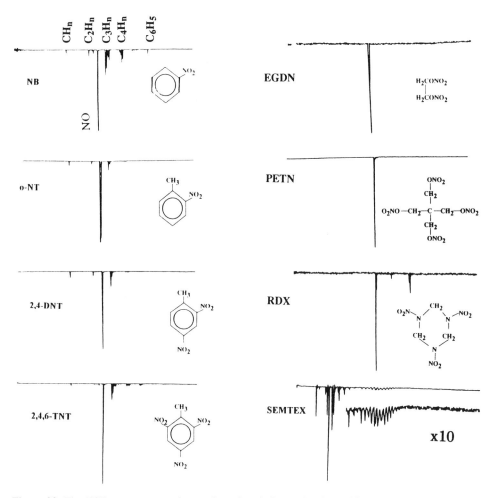

Figure 23. The TOF mass spectra of a number of explosive molecules at 226 nm. All spectra show a prominent mass peak at m/z = 30. To a degree the different types of explosives can be distinguised by the hydrocarbon fragments. Simeonssen et al. (1993) have shown that at 193 nm the fragmentation of the hydrocarbon clusters is much greater.

In a series of recent experiments at Glasgow and also in America, REMPI has been used to detect and distinguish nitroaromatic and explosive molecules with great sensitivity. All high explosive materials e.g. DNT, TNT, EGDN, PETN, RDX and SEMTEX (an RDX and PETN mixture) contain a number of NO_2 functional groups with the more dangerous explosives containing greater numbers of the group. It has been shown that these molecules can be fragmented with great efficiency to yield NO_x (x = 1,2) neutral molecules (Marshall et al., 1994; Simeonssen et al., 1993). On subsequent resonant multiphoton absorption of two or three photons at specific wavelengths, particularly 226 and 215 nm, these NO_x molecules can be ionized and detected with great sensitivity in a TOF. The very low vapour pressures from the explosives were admitted effusively to the acceleration region of the TOF mass spectrometer through a 0.5 mm diameter hole in the centre of the sample stub using a precision leak valve and capillary tube arrangement. The laser-sample interaction occurred at a distance of 1 mm from the sample stub and the ions passed into a 1.2 m field-free drift space where they were detected by a standard Thorn-EMI 18-dynode electron multiplier. The TOF mass spectrometer was operated in a Wiley-McLaren configuration with a mass resolution of 220 at m/z = 77.

The spectra for a number of explosives at 226.3 nm is shown in Fig. 23 with all the spectra showing a prominent peak at m/z = 30 (NO) characteristic of all nitro-explosive materials, and to some degree the different families of explosives can be distinguished by the other hydrocarbon fragments. It is possible that the 215 nm wavelength might provide a greater yield of hydrocarbon fragments and hence a greater degree of selectivity among the different explosive types (Simeonssen et al., 1993). The sensitivity levels for the detection of different types of explosives using laser mass spectrometry is typically between 10 –100 pg.

REMPI has been used in a similar way by Syage (1990) to detect the presence of organophosphonate compounds. These molecules are used extensively in chemical warfare agents and lasers have been used to fragment and ionize the molecules with great sensitivity, using the PO radical in this case as the fingerprint, in an identical way to the NO fragment in explosive materials.

To improve molecular selectivity, albeit at the expense of sensitivity, the molecular rotational degrees of freedom can be reduced to a few sharp atomic-like peaks by introducing the molecules into supersonic beams. Using the Joule-Kelvin effect, rapid cooling is realised by converting the energy of the internal vibrational and rotational degrees of freedom into translational energy via collisions. A system using jet cooling in conjunction with laser desorption and post-ionization is shown in Fig.24 (Grotemeyer and Schlag, 1989). An interesting comparison of the different techniques of exciting and ionizing a molecule is provided in Fig.25, where aniline spectra recorded under various conditions are presented. Jet-cooling sharpens the resonance features considerably (Marshall et al., 1991).

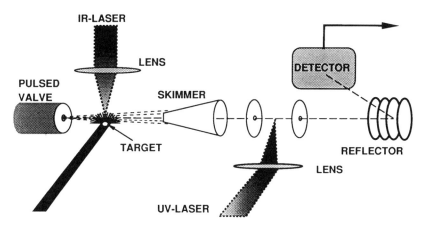

Figure 24. Sample irradiated by an IR laser to produce molecules that are seeded into a supersonic jet. These are post-ionized by a UV laser and then analysed by a reflectron mass spectrometer (Grotemeyer and Schlag, 1989).

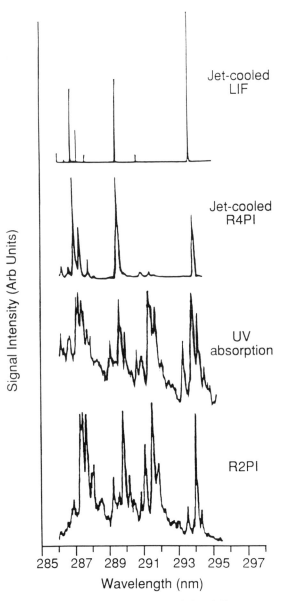

Figure 25. Analine spectrum recorded by jet cooled laser-induced fluorescence, jet-cooled R4PI (two photons to excite, two photons to ionize), single photon UV absorption and R2PI (Marshall et al., 1991). It is clear that the cooling considerably simplifies the spectrum.

With conventional analytical techniques, molecular isomers are extremely difficult to distinguish from each other. Although they have the same molecular weight, they have different absorption spectra and their ionization potentials may also be different. Supersonic jet-cooling is essential to reduce the rotational-vibrational structure, so that the distinguishing sharp absorption peaks result. Fig.26 shows the R2PI spectra for four isomers of dichlorotoluene. The wavelength spectra were obtained after expanding several ppm of each compound in a 1-atm reservoir of argon into vacuum at 10 torr (Tembruell et al., 1985).

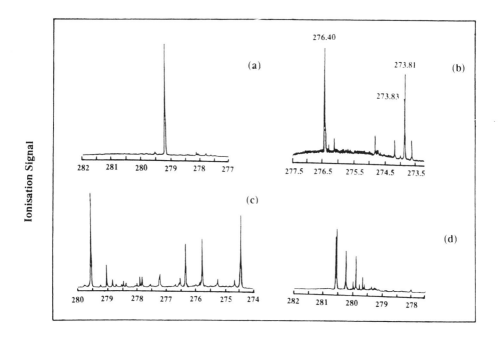

Wavelength (nm)

Figure 26. Resonant two-photon ionization spectrum of the isomers of dichlorotoluene taken in an expansion of 1 atm back-pressure of Ar of: (a) 2,4 dichlorotoluene, (b) 2,6 dichlorotoluene, (c) 2,5 dichlorotoluene and (d) 3,4 dichlorotoluene. Only the parent molecular ion is monitored in these spectra using a TOF mass spectrometer (Tembruell et al., 1985). (Reproduced by permission of the authors and the American Chemical Society).

CONCLUSIONS

Laser mass spectrometry has been shown to be a trace analytical technique of considerable potential, qualitatively and quantitatively, for both atoms and molecules. Already several commercial firms are advertising RIMS instruments of various types and in addition, laser microprobes can be fitted with a post-ablation ionization capability, both resonant and non-resonant. In the last two or three years, MALDI time of flight mass spectrometers have been developed by a number of companies and this is one of the real growth areas in scientific instrumentation. One of the difficulties which instrument manufacturers identify with RIMS technology is the use of the dye lasers with their inherent bulkiness and operating difficulties. With the development of solid state tunable lasers (Ti-sapphire and optical parametric oscillators) as well as semiconductor diode lasers it is likely that RIMS will soon reach its full potential and be accorded similar acceptance as SIMS (Ledingham and Singhal, 1991). One of the exciting new developments in laser mass spectrometry especially for REMPI is the development of the femtosecond laser systems. The lifetime of the dissociative states of molecules is typically femtoseconds and hence if laser pulses of widths less than these lifetimes are used the fragmentation processes can be largely defeated, resulting in large parent mass signals being detected.

ACKNOWLEDGEMENTS

It is a pleasure to acknowledge the staff and research students of the Glasgow LIS group without whose hard work the contents of this paper would not have been possible.

REFERENCES

Ambartzumian, R.V., and Letokhov, V.S., 1972, *Appl.Opt.* 11:354.

Arlinghaus, H.F., Spaar, M.T., and Thonnard, N., 1990, *J. Vac. Sci. Technol.* A8:2318.

Arlinghaus, H.F., Thonnard, N., Sparr, M.T., Sachleben, R.A., Brown, G.M., Foote, R.S., Sloop, F.V., Petersen, J.R., and Jacobsen, K.B., 1991, *J. Vac. Sci. Technol* A9:1312.

Becker, C.H., and Gillen, K.T., 1984, *Anal. Chem.* 56:1671.

Bekov, G.I., and Letokhov, V.S., 1983, *Appl. Phys.* B30:161.

Bekov, G.I., Letokhov, V.S., and Radaev, V.N., 1985, *J. Opt. Soc. Am.* B 2:1554.

Boesl, U., Neusser, H.J., and Schlag, E.W., 1981, *Chem. Phys.* 55:193.

Boesl, U., 1991, *J. Phys. Chem.* 95:2949.

Borthwick, I.S., Ledingham,K.W.D., and Singhal, R.P., 1992, *Spectrochimica Acta B* 47B:1259.

Bushaw, B.A., 1992, *in:* "Resonance Ionization Spectroscopy 92," Inst. Phys. Con. Ser. 128, I.O.P., Bristol.

Calaway, W.F., Coon, S.R., Pellin, M.J., Young, C.E., Whitten, J.E., Wiens, R.C., Gruen, D.M., Stingeder, G., Penka, V., Grasserbauer, M., and Burnett, D.S., 1992, *in:* "Resonance Ionization Spectroscopy 92," Inst. Phys. Con. Ser. 128, I.O.P., Bristol.

Clarke, N.S., 1989, *Chemistry in Britain* 25(5):484.

Dominic Chan, T-W., Colburn, A.W., and Derrick, P.J., 1992, *Org. Mass Spectrom.* 27:53.

Downey, S.W., Emerson, A.B., Kopf, R.F., and Kuo, J.M., 1990, *Surf. Int. Anal.* 15:781.

Downey, S.W., and Emerson, A.B., 1991, *Anal.Chem.* 63:916.

Eiden, G.C., Anderson, J.E., and Nogar, N.S., *to be published.*

Fenner, N.C., and Daly, N.R., 1966, *Rev. Sci. Instr.* 37:1068.

Göppert-Mayer, M., 1931, *Ann. Phys. LPZ.* 9:273.

Grotemeyer, J., and Schlag, E.W., 1989, *Acc. Chem. Res.* 22:399.

Hillenkamp, F., Karas, M., Beavis, R.C., and Chait, B.T., 1991, *Anal. Chem.* 63:1193A.

Hillenkamp, F., 1994, *private communication*

Hurst, G.S., Payne, M.G., Kramer, S.D., and Young, J.P., 1979, *Rev. Mod. Phys.* 51:767.

Hurst, G.S., and Payne, M.G., 1988, "Principles and Applications of Resonance Ionization Spectroscopy," Adam Hilger, Bristol and Philadelphia.

Karas, M., Bachmann, D., Bahr, U., and Hillenkamp, F., 1987, *Int. J. Mass Spectrom. Ion Proc.* 78:53.

Ledingham, K.W.D., and Singhal, R.P., 1991, *J. Anal. Atom. Spectrom.* 6:73.

Ledingham, K.W.D., and Singhal, R.P., 1992, Laser mass spectrometry, *in:* Applied Laser Spectroscopy - Techniques Instrumentation and Applications," D.L. Andrews, ed., VCH Publishers Inc., New York.

Letokhov, V.S., 1987, "Laser Photoionization Spectroscopy," Academic Press, London.

Lubman, D.M., ed., 1990, "Lasers and Mass Spectrometry," Oxford University Press, New York.

McCombes, P.T., Borthwick, I.S., Jennings, R., Ledingham, K.W.D., and Singhal, R.P., 1991, *in:* "Optogalvanic Spectroscopy," R S Stewart and J E Lawler, ed., Inst. Phys. Conf. Ser.113, I.O.P., Bristol.

Marshall, A., Clark, A., Jennings, R., Ledingham, K.W.D., and Singhal, R.P., 1991, *Meas. Sci. Technol.* 2:1078.

Marshall, A., Clark, A., Ledingham, K.W.D., Sander, J., Singhal, R.P., Kosmidis, C., and Deas, R.M., 1994, *Rapid Commun. Mass Spectrom.* 8:521.

Olaf-Bornsen, K., Schar, M., and Widmer, H.M., 1990, *Chimica* 44:412.

Parks, J.E., Beekman, D.W., Moore, L.J., Schmitt, H.W., Spaar, M.T., Taylor, E.H., Hutchinson, J.M.R., and Fairbank, W.M., Jr, 1986, *in:* "Resonance Ionization Spectroscopy 86," Inst. Phys. Conf. Ser. 84, I.O.P., Bristol.

Payne, M.G., Deng Lu, and Thonnard, N., 1994, *Rev. Sci. Instrum.* 65:2433

Scott, C.T.J., Kosmidis, C., Jia, W.J., Ledingham, K.W.D., and Singhal, R.P., 1994, *Rapid Comm. Mass Spectrom.* 8:829.

Simeonssen, J.B., Lemire, G.W., and Sausa, R.C., 1993, *Appl. Spec.* 47:1907.

Singhal, R.P., Land, A.P., Ledingham, K.W.D., and Towrie, M., 1989, *J. Anal. Atom. Spectrom.* 4:599.

Syage, J.A., 1990, *Anal. Chem.* 62:505A.

Tanaka,K., Ido,Y., Akita, S., Yoshida, Y., and Yoshida, T., 1987, *in:* "Proceediings of the Second Japan-China Joint Symposium on Mass Spectrometry," Osaka, Japan.

Tanaka, K., Waki, H., Ido, W., Akita, S., Yoshida, Y., and Yoshida, T., 1988, *Rapid Commun. Mass Spectrom.* 2:151.

Tembruell, R., Sin, C.H., Li, P., Pang, H.M., and Lubman, D.M., 1985, *Anal. Chem.* 57:1186.

Thonnard, N., Parks, J.E., Willis, R.D., Moore, L.J., and Arlinghaus, H.F., 1989, *Surf. Int. Anal.* 14:751.

Thonnard, N., Wright, M.C., Davis, W.A., and Willis, R.D., 1992, *in:* "Resonance Ionization Spectroscopy 92," Inst. Phys. Conf. Ser. 128, I.O.P., Bristol.

Towrie, M., Drysdale, S.L.T., Jennings, R., Land, A.P., Ledingham, K.W.D., McCombes, P.T., Singhal, R.P., Smyth, M.H.C., and McLean, C.J., 1990, *Int. J. of Mass Spec. and Ion Proc.* 26:309.

Urban, F.J., Deissenberger, R., Hermann, G., Kohler, S., Riegel, J., Trautmann, N., Wendeler, H., Albus, F., Ames, F., Kluge, H-J., Krass, S., and Scheerer, F., 1992, *in:* "Resonance Ionization Spectroscopy 92," Inst. Phys. Conf. Ser. 128, I.O.P., Bristol.

Vertes, A., Gijbels, R., and Adams, F., eds., 1993, "Laser Ionization Mass Analysis," *Chemical Analysis,* Ser. Vol. 124, John Wiley & Sons, Inc., New York.

Wang, L., Ledingham, K.W.D., McLean, C.J., and Singhal, R.P., 1992, *Appl. Phys.* B54:71.

Wendt, K., Passler, G., and Trautmann, N., *to be published*

INDEX

Quantum cascade laser, 15
Quantum detectors, *see* Photodiode,
　　　Photomultiplier tube
Quantum efficiency, 22
Quantum well, 15
Quantum yield, *see* Fluorescence quantum yield

Radiant sensitivity, 24
Radiative lifetime, *see* Lifetime, radiative
Radiative relaxation, *see* Spontaneous emission
Radioisotope detection, 216-218
Raman spectroscopy, 91-130; *see also* Coherent
　　　anti-Stokes Raman spectroscopy,
　　　Stimulated Raman scattering
　Fourier transform, 106, 116, 126
　imaging and microprobe, 109-111, 116, 119,
　　　123-124
　lidar, 137, 138-139, 143
　optical fibre, 111-112
　resonance, 95-96, 104, 125, 126
　spin-flip, 102, 103
　surface-enhanced, 112-113
Rayleigh length, 157, 168
Recoil anisotropy, 59
Remote sensing, 133-145
Resolving power, 30
Resonance enhanced multiphoton ionisation, 49-50,
　　　55, 222, 223, 225; *see also* Resonance
　　　ionisation mass spectrometry
Resonance ionisation mass spectrometry, 199-211
Resonant laser ablation, 212-216
Rhodamine B, 4
Rice flour, 106
Ring dye lasers, *see* Dye lasers, ring
Rubber, 119-120
Ruby laser, 1, 8-9, 149

Saturation, 24, 52, 204, 205
　broadening, *see* Line broadening, power
Secondary ionisation mass spectrometry, 199-200,
　　　210
Second harmonic generation, 10, 151, 152, 153-
　　　159, 171, 172-184
Self-focusing, 165
Self phase modulation, 165-166
Semiconductors, 101-104, 213-214, 216, 218-220
Shot noise, 81, 127
Spiking, 20

Spin-flip Raman scattering, *see* Raman scattering,
　　　spin-flip
Spontaneous emission, 3, 35, 36, 105-106
Sputter-initiated resonance ionisation spectrometry,
　　　210, 220
Stark tuning, 14, 44-45
Steel, 215-216
Stimulated emission, 35　pumping, 55-58
Stimulated Raman scattering, 12-13, 116
Sub-Doppler spectroscopy, 52-54
Sum-frequency generation, 10, 151, 159-160, 171,
　　　172-174, 180-181
Surface-enhanced Raman spectroscopy, *see* Raman
　　　spectroscopy, surface-enhanced
Surface non-linear optics, 171-184

Temperature determination, 50-51, 138-139, 143
Terephthalic acid, 122
Thermal detectors, 26-28
Thermal lensing spectroscopy, 45-47
Three-level laser, 4, 6
Titanium-sapphire laser, 9, 10, 152, 225
Toluene, $C_6H_5CH_3$, 221-222
Trace analysis, 50, 87, 194, 205, 208-210, 211,
　　　216-218, 220-225
Transform limit, 7-8, 75-76
Transition metal ion vibronic lasers, 8-10
Two-photon absorption, 52-54, 193, 199, 201-202,
　　　203-204, 220-221, 224, 225

Ultrafast spectroscopy, *see* Femtosecond lasers,
　　　Picosecond measurements

Vegetation sensing, 143-145
Velocity modulation, 66-67, 72
Vibrational redistribution, intramolecular, 40
Vibrational relaxation, *see* Collision-induced
　　　vibrational relaxation
Vibronic lasers, *see* Dye lasers, Transition metal
　　　ion vibronic lasers

Water, 143
White cell, *see* Multipass cell

Xenon chloride laser, 161, 221

Zeeman effect, 14, 43-44, 63, 103, 104
Zinc telluride, ZnTe, 101-102